Advanced Electrical Circuit Analysis

Mehdi Rahmani-Andebili

Advanced Electrical Circuit Analysis

Practice Problems, Methods, and Solutions

 Springer

Mehdi Rahmani-Andebili
Engineering Technology
State University of New York
Buffalo, NY, USA

ISBN 978-3-030-78542-0   ISBN 978-3-030-78540-6   (eBook)
https://doi.org/10.1007/978-3-030-78540-6

This Springer imprint is published by the registered company Springer Nature Switzerland AGThe registered company
address is: Gewerbestrasse 11, 6330 Cham, Switzerland

Preface

Electrical circuit analysis is one of the most fundamental subjects of electrical engineering major. This textbook includes the advanced subjects of electrical circuit analysis that have not been covered in the previously published textbooks, that is, *DC Electrical Circuit Analysis* and *AC Electrical Circuit Analysis*. The subjects include state equations of electrical circuits, Laplace transform and network function, natural frequencies of electrical circuits, network theorems (Tellegen's and linear time-invariant network theorems), and two-port networks.

Like the previously published textbooks, this textbook includes very detailed and multiple methods of problem solutions. It can be used as a practicing textbook by students and as a supplementary teaching source by instructors.

To help students study the textbook in the most efficient way, the exercises have been categorized in nine different levels. In this regard, for each problem of the textbook, a difficulty level (easy, normal, or hard) and a calculation amount (small, normal, or large) have been assigned. Moreover, in each chapter, problems have been ordered from the easiest problem with the smallest calculations to the most difficult problem with the largest calculations. Therefore, students are advised to study the textbook from the easiest problems and continue practicing till they reach the normal and then the hardest ones. On the other hand, this classification can help instructors choose their desirable problems to conduct a quiz or a test. Moreover, the classification of computation amount can help students manage their time during future exams and instructors give the appropriate problems based on the exam duration.

Since the problems have very detailed solutions and some of them include multiple methods of solution, the textbook can be useful for the under-prepared students. In addition, the textbook is beneficial for knowledgeable students because it includes advanced exercises.

In the preparation of problem solutions, an attempt has been made to use typical methods of electrical circuit analysis to present the textbook as an instructor-recommended one. In other words, the heuristic methods of problem solution have never been used as the first method of problem solution. By considering this key point, the textbook will be in the direction of instructors' lectures, and the instructors will not see any untaught problem solutions in their students' answer sheets.

The Iranian University Entrance Exam for the master's and PhD degrees of electrical engineering major is the main reference of the textbook; however, all the problem solutions have been provided by me. The Iranian University Entrance Exam is one of the most competitive university entrance exams in the world that allows only 10% of the applicants to get into prestigious and tuition-free Iranian universities.

Buffalo, NY, USA   Mehdi Rahmani-Andebili

Contents

About the Author

Mehdi Rahmani-Andebili is an assistant professor in the Department of Engineering Technology at State University of New York, Buffalo State. He received his first M.Sc. and Ph.D. degrees in electrical engineering (power system) from Tarbiat Modares University and Clemson University in 2011 and 2016, respectively, and his second M.Sc. degree in physics and astronomy from the University of Alabama in Huntsville in 2019. Moreover, he was a postdoctoral fellow at Sharif University of Technology during 2016–2017. As a professor, he has taught many courses such as Essentials of Electrical Engineering Technology, Electrical Circuits Analysis I, Electrical Circuits Analysis II, Electrical Circuits and Devices, Industrial Electronics, and Renewable Distributed Generation and Storage. Dr. Rahmani-Andebili has more than hundred single-author publications including textbooks, books, book chapters, journal papers, and conference papers. His research areas include smart grid, power system operation and planning, integration of renewables and energy storages into power system, energy scheduling and demand-side management, plug-in electric vehicles, distributed generation, and advanced optimization techniques in power system studies.

Abstract

In this chapter, state equations are applied to solve the basic and advanced problems of electrical circuit analysis. In this chapter, the problems are categorized in different levels based on their difficulty levels (easy, normal, and hard) and calculation amounts (small, normal, and large). Additionally, the problems are ordered from the easiest problem with the smallest computations to the most difficult problems with the largest calculations.

1.1. In the circuit of Figure 1.1, $v_C(t)$ and $i_L(t)$ are the state variables [1–2]. Write the output voltage ($v_o(t)$) based on the state variables.

Difficulty level   ○ Easy   ● Normal   ○ Hard
Calculation amount   ○ Small   ● Normal   ○ Large

1. $i_L(t) + \frac{1}{6}v_C(t) - \frac{1}{2}v_s(t)$
2. $6i_L(t) - v_C(t) - v_s(t)$
3. $3i_L(t) - \frac{1}{2}v_C(t) - \frac{1}{2}v_s(t)$
4. $2i_L(t) + \frac{1}{3}v_C(t) - v_s(t)$

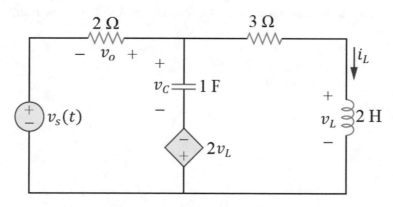

Figure 1.1 The circuit of problem 1.1

1.2. In the circuit of Figure 1.2, if $\mathbf{X} = \begin{bmatrix} i_L(t) \\ v_C(t) \end{bmatrix}$ is assigned as the state vector, determine matrices **A** and **B** in the relation of

$$\dot{\mathbf{X}} = \mathbf{A}\mathbf{X} + \mathbf{B}\begin{bmatrix} i_s(t) \\ v_s(t) \end{bmatrix}.$$

Difficulty level   ○ Easy   ● Normal   ○ Hard
Calculation amount   ○ Small   ● Normal   ○ Large

1. $\mathbf{A} = \begin{bmatrix} 1 & -1 \\ -1 & -1 \end{bmatrix}, \mathbf{B} = \begin{bmatrix} 1 & 0 \\ -1 & 1 \end{bmatrix}$

2. $\mathbf{A} = \begin{bmatrix} -1 & 1 \\ 1 & -1 \end{bmatrix}, \mathbf{B} = \begin{bmatrix} 1 & 0 \\ -1 & 1 \end{bmatrix}$

3. $\mathbf{A} = \begin{bmatrix} -1 & -1 \\ 1 & -1 \end{bmatrix}, \mathbf{B} = \begin{bmatrix} 1 & 0 \\ -1 & 1 \end{bmatrix}$

4. $\mathbf{A} = \begin{bmatrix} -1 & -1 \\ 1 & -1 \end{bmatrix}, \mathbf{B} = \begin{bmatrix} 0 & 1 \\ -1 & 1 \end{bmatrix}$

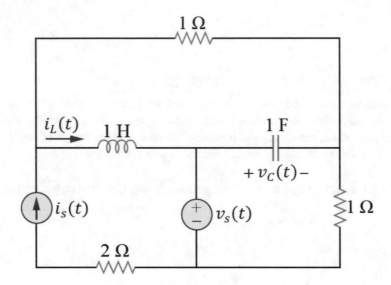

Figure 1.2 The circuit of problem 1.2

1.3. In the circuit of Figure 1.3, the state equations are in the form of $\dot{\mathbf{X}} = \mathbf{AX}$. Determine matrix $\mathbf{A}$, for the state vector of $\mathbf{X} = \begin{bmatrix} i_L(t) \\ v_C(t) \end{bmatrix}$.

Difficulty level   ○ Easy   ● Normal   ○ Hard
Calculation amount   ○ Small   ● Normal   ○ Large

1. $\mathbf{A} = \begin{bmatrix} -\dfrac{8}{3} & -2 \\ 1 & 0 \end{bmatrix}$

2. $\mathbf{A} = \begin{bmatrix} -\dfrac{8}{3} & 2 \\ -1 & 0 \end{bmatrix}$

3. $\mathbf{A} = \begin{bmatrix} -\dfrac{4}{3} & 2 \\ 1 & 0 \end{bmatrix}$

4. $\mathbf{A} = \begin{bmatrix} -\dfrac{4}{3} & 2 \\ 0 & -1 \end{bmatrix}$

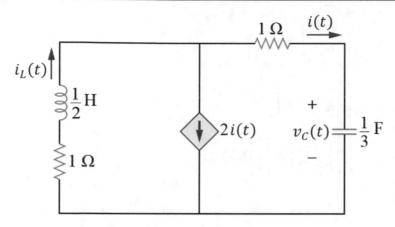

Figure 1.3 The circuit of problem 1.3

1.4. Determine matrix **A** of the state equations ($\dot{\mathbf{X}} = \mathbf{A}\mathbf{X} + \mathbf{B}w$) for the circuit of Figure 1.4 if $\mathbf{X} = \begin{bmatrix} v_C(t) \\ i_L(t) \end{bmatrix}$.

Difficulty level   ○ Easy   ● Normal   ○ Hard
Calculation amount   ○ Small   ● Normal   ○ Large

1. $A = \begin{bmatrix} -1 & 1 \\ -1 & 3 \end{bmatrix}$

2. $A = \begin{bmatrix} 1 & 3 \\ 1 & 1 \end{bmatrix}$

3. $A = \begin{bmatrix} -0.5 & 0.5 \\ -0.5 & -1.5 \end{bmatrix}$

4. $A = \begin{bmatrix} 0.5 & 1.5 \\ 0.5 & 0.5 \end{bmatrix}$

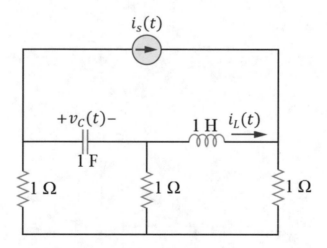

Figure 1.4 The circuit of problem 1.4

1.5. Determine the state equations of the circuit of Figure 1.5 based on the given voltage of the capacitor and current of the inductor.

Difficulty level   ○ Easy   ● Normal   ○ Hard
Calculation amount   ○ Small   ● Normal   ○ Large

1. $\begin{bmatrix} \dfrac{d}{dt}v_C(t) \\ \dfrac{d}{dt}i_L(t) \end{bmatrix} = \begin{bmatrix} \dfrac{3}{2} & 1 \\ -\dfrac{1}{2} & 0 \end{bmatrix} \begin{bmatrix} v_C(t) \\ i_L(t) \end{bmatrix} + \begin{bmatrix} 1 \\ -\dfrac{1}{2} \end{bmatrix} v_s(t)$

2. $\begin{bmatrix} \dfrac{d}{dt} v_C(t) \\ \dfrac{d}{dt} i_L(t) \end{bmatrix} = \begin{bmatrix} -\dfrac{3}{2} & 1 \\ -\dfrac{1}{2} & 0 \end{bmatrix} \begin{bmatrix} v_C(t) \\ i_L(t) \end{bmatrix} + \begin{bmatrix} 1 \\ -\dfrac{1}{2} \end{bmatrix} v_s(t)$

3. $\begin{bmatrix} \dfrac{d}{dt} v_C(t) \\ \dfrac{d}{dt} i_L(t) \end{bmatrix} = \begin{bmatrix} -\dfrac{3}{2} & 1 \\ -\dfrac{1}{2} & 0 \end{bmatrix} \begin{bmatrix} v_C(t) \\ i_L(t) \end{bmatrix} + \begin{bmatrix} 1 \\ \dfrac{1}{2} \end{bmatrix} v_s(t)$

4. $\begin{bmatrix} \dfrac{d}{dt} v_C(t) \\ \dfrac{d}{dt} i_L(t) \end{bmatrix} = \begin{bmatrix} -\dfrac{3}{2} & -1 \\ -\dfrac{1}{2} & 0 \end{bmatrix} \begin{bmatrix} v_C(t) \\ i_L(t) \end{bmatrix} + \begin{bmatrix} 1 \\ \dfrac{1}{2} \end{bmatrix} v_s(t)$

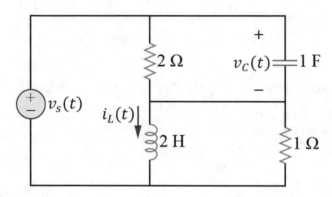

Figure 1.5 The circuit of problem 1.5

1.6. In the circuit of Figure 1.6, determine matrix $\mathbf{A}$ if the state equations are in the form of $\dot{\mathbf{X}} = \mathbf{A}\mathbf{X}$.

Difficulty level   ○ Easy   ● Normal   ○ Hard
Calculation amount   ○ Small   ● Normal   ○ Large

1. $\mathbf{A} = \begin{bmatrix} 0 & 2 \\ -1 & 1 \end{bmatrix}$

2. $\mathbf{A} = \begin{bmatrix} 1 & 1 \\ -2 & 0 \end{bmatrix}$

3. $\mathbf{A} = \begin{bmatrix} 1 & 0 & 1 \\ -1 & 0 & 0 \\ 1 & 1 & -1 \end{bmatrix}$

4. $\mathbf{A} = \begin{bmatrix} 2 & 2 \\ -1 & 0 \end{bmatrix}$

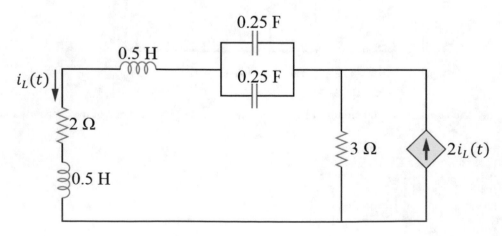

Figure 1.6 The circuit of problem 1.6

1.7. If the state equations of the circuit of Figure 1.7 are presented in the form of $\dot{\mathbf{X}} = \mathbf{A}\mathbf{X} + \mathbf{B}i_s(t)$, determine vector **B**, where

$$\mathbf{X} = \begin{bmatrix} i_L(t) \\ v_C(t) \end{bmatrix}.$$

Difficulty level   ○ Easy   ● Normal   ○ Hard
Calculation amount   ○ Small   ○ Normal   ● Large

1. $\mathbf{B} = \begin{bmatrix} 1 \\ 1 \end{bmatrix}$

2. $\mathbf{B} = \begin{bmatrix} 1 \\ -1 \end{bmatrix}$

3. $\mathbf{B} = \begin{bmatrix} 0 \\ 1 \end{bmatrix}$

4. $\mathbf{B} = \begin{bmatrix} 1 \\ 0 \end{bmatrix}$

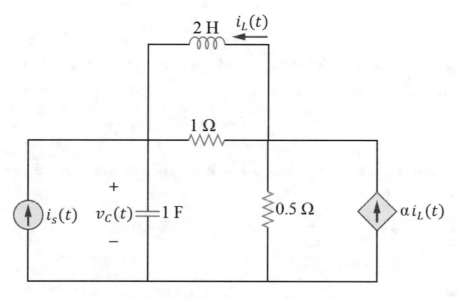

Figure 1.7 The circuit of problem 1.7

1.8. In the circuit of Figure 1.8, by choosing $\mathbf{X} = \begin{bmatrix} v_C(t) \\ i_L(t) \end{bmatrix}$ as the state vector, what is matrix **A** in the state equations of

$\dot{\mathbf{X}} = \mathbf{A}\mathbf{X} + \mathbf{B}W$?

Difficulty level   ○ Easy   ● Normal   ○ Hard
Calculation amount   ○ Small   ○ Normal   ● Large

1. $\mathbf{A} = \begin{bmatrix} -1 & 1 \\ -1 & -2 \end{bmatrix}$

2. $\mathbf{A} = \begin{bmatrix} \dfrac{1}{3} & -\dfrac{1}{3} \\ \dfrac{1}{3} & \dfrac{2}{3} \end{bmatrix}$

3. $\mathbf{A} = \begin{bmatrix} -\dfrac{1}{3} & \dfrac{1}{3} \\ -\dfrac{1}{3} & -\dfrac{2}{3} \end{bmatrix}$

4. $\mathbf{A} = \begin{bmatrix} 1 & -1 \\ 1 & 2 \end{bmatrix}$

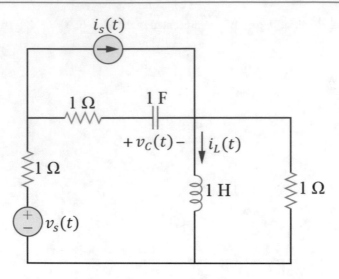

Figure 1.8 The circuit of problem 1.8

1.9. A network includes some resistors, an inductor of 1 H, and a capacitor of 1 F. By choosing the state vector of $\begin{bmatrix} i_L(t) \\ v_C(t) \end{bmatrix}$, the system matrix of the state equations is as follows:

$$\mathbf{A} = \begin{bmatrix} a_{11} & a_{12} \\ a_{21} & a_{22} \end{bmatrix}$$

Determine the updated system matrix of the state equations if the place of the inductor and the capacitor is changed and $\begin{bmatrix} v_C(t) \\ i_L(t) \end{bmatrix}$ is chosen as the new state vector.

Difficulty level   ○ Easy   ○ Normal   ● Hard
Calculation amount   ○ Small   ○ Normal   ● Large

1. It is impossible to determine $\mathbf{A_{new}}$.

2. $\mathbf{A_{new}} = \frac{1}{Det[\mathbf{A}]} \begin{bmatrix} a_{22} & -a_{12} \\ -a_{21} & a_{11} \end{bmatrix}$.

3. $\mathbf{A_{new}} = \frac{1}{Det[\mathbf{A}]} \begin{bmatrix} a_{11} & -a_{21} \\ -a_{12} & a_{22} \end{bmatrix}$.

4. $\mathbf{A_{new}} = \frac{1}{Det[\mathbf{A}]} \begin{bmatrix} \dfrac{1}{a_{22}} & -\dfrac{1}{a_{12}} \\ -\dfrac{1}{a_{21}} & \dfrac{1}{a_{11}} \end{bmatrix}$.

1.10. In the circuit of Figure 1.9, the state vector and the input vector are $\mathbf{X} = \begin{bmatrix} i_{L1}(t) \\ i_{L2}(t) \\ v_C(t) \end{bmatrix}$ and $\mathbf{W} = \begin{bmatrix} i_s(t) \\ v_s(t) \end{bmatrix}$, respectively. If the state equations of the circuit are written in the form of $\dot{\mathbf{X}} = \mathbf{AX} + \mathbf{B}\begin{bmatrix} i_s(t) \\ v_s(t) \end{bmatrix}$, determine matrix $\mathbf{B}$, while we have:

$$\mathbf{A} = \begin{bmatrix} -1 & -1 & 0 \\ -\dfrac{1}{2} & -\dfrac{1}{2} & -\dfrac{1}{2} \\ 0 & 1 & 0 \end{bmatrix}$$

Difficulty level   ○ Easy   ○ Normal   ● Hard
Calculation amount   ○ Small   ○ Normal   ● Large

1. $\mathbf{B} = \begin{bmatrix} -1 & 1 \\ -\dfrac{1}{2} & \dfrac{1}{2} \\ -1 & 0 \end{bmatrix}$

2. $\mathbf{B} = \begin{bmatrix} 1 & 1 \\ -\dfrac{1}{2} & -\dfrac{1}{2} \\ 0 & -1 \end{bmatrix}$

3. $\mathbf{B} = \begin{bmatrix} -1 & -1 \\ \dfrac{1}{2} & \dfrac{1}{2} \\ 1 & 0 \end{bmatrix}$

4. $\mathbf{B} = \begin{bmatrix} 1 & 1 \\ \dfrac{1}{2} & \dfrac{1}{2} \\ -1 & 0 \end{bmatrix}$

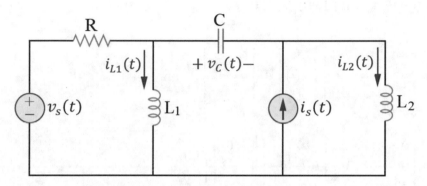

Figure 1.9 The circuit of problem 1.10

1.11. In the circuit of Figure 1.10, $\mathbf{X} = \begin{bmatrix} v_C(t) \\ i_L(t) \end{bmatrix}$ is the state vector. Determine matrix $\mathbf{A}$ if the state equations are written in the form of $\dot{\mathbf{X}} = \mathbf{A}\mathbf{X} + \mathbf{B}i_s(t)$.

Difficulty level   ○ Easy   ○ Normal   ● Hard
Calculation amount   ○ Small   ○ Normal   ● Large

1. $\mathbf{A} = \begin{bmatrix} -1 & -0.5 \\ 0.5 & 1.5 \end{bmatrix}$

2. $\mathbf{A} = \begin{bmatrix} -1 & -1 \\ -1 & 3 \end{bmatrix}$

3. $\mathbf{A} = \begin{bmatrix} -0.5 & -0.5 \\ 0.5 & -1.5 \end{bmatrix}$

4. $\mathbf{A} = \begin{bmatrix} -1 & -1 \\ 1 & -3 \end{bmatrix}$

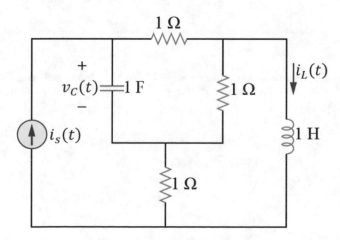

Figure 1.10 The circuit of problem 1.11

1.12. Determine the state equations of the circuit of Figure 1.11.

Difficulty level   ○ Easy   ○ Normal   ● Hard
Calculation amount   ○ Small   ○ Normal   ● Large

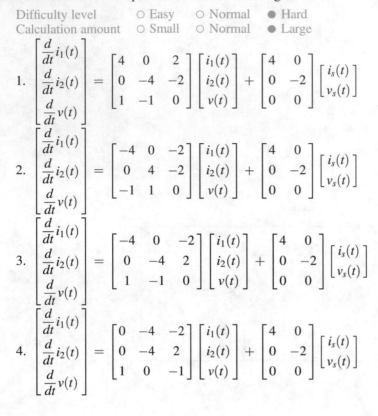

1. $\begin{bmatrix} \frac{d}{dt}i_1(t) \\ \frac{d}{dt}i_2(t) \\ \frac{d}{dt}v(t) \end{bmatrix} = \begin{bmatrix} 4 & 0 & 2 \\ 0 & -4 & -2 \\ 1 & -1 & 0 \end{bmatrix} \begin{bmatrix} i_1(t) \\ i_2(t) \\ v(t) \end{bmatrix} + \begin{bmatrix} 4 & 0 \\ 0 & -2 \\ 0 & 0 \end{bmatrix} \begin{bmatrix} i_s(t) \\ v_s(t) \end{bmatrix}$

2. $\begin{bmatrix} \frac{d}{dt}i_1(t) \\ \frac{d}{dt}i_2(t) \\ \frac{d}{dt}v(t) \end{bmatrix} = \begin{bmatrix} -4 & 0 & -2 \\ 0 & 4 & -2 \\ -1 & 1 & 0 \end{bmatrix} \begin{bmatrix} i_1(t) \\ i_2(t) \\ v(t) \end{bmatrix} + \begin{bmatrix} 4 & 0 \\ 0 & -2 \\ 0 & 0 \end{bmatrix} \begin{bmatrix} i_s(t) \\ v_s(t) \end{bmatrix}$

3. $\begin{bmatrix} \frac{d}{dt}i_1(t) \\ \frac{d}{dt}i_2(t) \\ \frac{d}{dt}v(t) \end{bmatrix} = \begin{bmatrix} -4 & 0 & -2 \\ 0 & -4 & 2 \\ 1 & -1 & 0 \end{bmatrix} \begin{bmatrix} i_1(t) \\ i_2(t) \\ v(t) \end{bmatrix} + \begin{bmatrix} 4 & 0 \\ 0 & -2 \\ 0 & 0 \end{bmatrix} \begin{bmatrix} i_s(t) \\ v_s(t) \end{bmatrix}$

4. $\begin{bmatrix} \frac{d}{dt}i_1(t) \\ \frac{d}{dt}i_2(t) \\ \frac{d}{dt}v(t) \end{bmatrix} = \begin{bmatrix} 0 & -4 & -2 \\ 0 & -4 & 2 \\ 1 & 0 & -1 \end{bmatrix} \begin{bmatrix} i_1(t) \\ i_2(t) \\ v(t) \end{bmatrix} + \begin{bmatrix} 4 & 0 \\ 0 & -2 \\ 0 & 0 \end{bmatrix} \begin{bmatrix} i_s(t) \\ v_s(t) \end{bmatrix}$

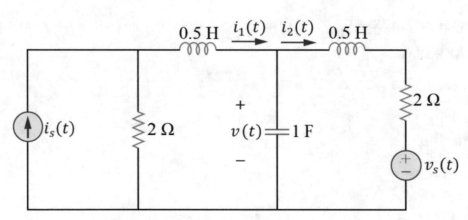

Figure 1.11 The circuit of problem 1.12

References

1. Rahmani-Andebili, M. (2020). DC Electrical circuit analysis: Practice problems, methods, and solutions, Springer Nature.
2. Rahmani-Andebili, M. (2020). AC Electrical circuit analysis: Practice problems, methods, and solutions, Springer Nature.

Solutions of Problems: State Equations of Electrical Circuits

2

Abstract

In this chapter, the problems of the first chapter are fully solved, in detail, step-by-step, and with different methods.

2.1. Applying KVL in the right-side mesh [1–2]:

$$2v_L(t) - v_C(t) + 3i_L(t) + v_L(t) = 0 \Rightarrow 3v_L(t) - v_C(t) + 3i_L(t) = 0 \Rightarrow v_L(t) = \frac{1}{3}v_C(t) - i_L(t) \tag{1}$$

Applying KVL in the left-side mesh:

$$-v_s(t) - v_o(t) + v_C(t) - 2v_L(t) = 0 \overset{(1)}{\Rightarrow} v_o(t) = -v_s(t) + v_C(t) - 2\left(\frac{1}{3}v_C(t) - i_L(t)\right)$$

$$\Rightarrow v_o(t) = 2i_L(t) + \frac{1}{3}v_C(t) - v_s(t)$$

Choice (4) is the answer.

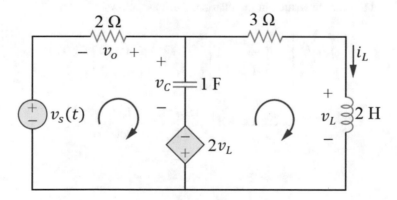

Figure 2.1 The circuit of solution of problem 2.1

2.2. Applying KCL in node 1 in the circuit of Figure 2.2.2:

$$-i_s(t) + i_L(t) + i_1(t) = 0 \Rightarrow i_1(t) = i_s(t) - i_L(t) \tag{1}$$

Applying KCL in node 2 in the circuit of Figure 2.2.2:

$$-i_1(t) - i_C(t) + i_2(t) = 0 \Rightarrow i_2(t) = i_1(t) + i_C(t) \overset{(1)}{\Rightarrow} i_2(t) = i_s(t) - i_L(t) + i_C(t) \tag{2}$$

As we know, the current-voltage relation of inductor and the voltage-current relation of capacitor are as follows:

$$i_C(t) = C\frac{d}{dt}v_C(t) \overset{C=1}{\Longrightarrow} i_C(t) = \frac{d}{dt}v_C(t) \tag{3}$$

$$v_L(t) = L\frac{d}{dt}i_L(t) \overset{L=1}{\Longrightarrow} v_L(t) = \frac{d}{dt}i_L(t) \tag{4}$$

Solving (2) and (3):

$$i_2(t) = i_s(t) - i_L(t) + \frac{d}{dt}v_C(t) \tag{5}$$

Applying KVL in the top mesh:

$$-v_L(t) + i_1(t) - v_C(t) = 0 \Rightarrow v_L(t) = i_1(t) - v_C(t) \overset{(1),(4)}{\Longrightarrow} \frac{d}{dt}i_L(t) = -i_L(t) - v_C(t) + i_s(t) \tag{6}$$

Applying KVL in the lower-right mesh:

$$-v_s(t) + v_C(t) + i_2(t) = 0 \overset{(2)}{\Rightarrow} -v_s(t) + v_C(t) + i_s(t) - i_L(t) + i_C(t) = 0$$

$$\overset{(3)}{\Rightarrow} -v_s(t) + v_C(t) + i_s(t) - i_L(t) + \frac{d}{dt}v_C(t) = 0 \Rightarrow \frac{d}{dt}v_C(t) = i_L(t) - v_C(t) - i_s(t) + v_s(t) \tag{7}$$

The equations of (6) and (7) can be written in the matrices form as follows:

$$\begin{bmatrix} \frac{d}{dt}i_L(t) \\ \frac{d}{dt}v_C(t) \end{bmatrix} = \begin{bmatrix} -1 & -1 \\ 1 & -1 \end{bmatrix} \begin{bmatrix} i_L(t) \\ v_C(t) \end{bmatrix} + \begin{bmatrix} 1 & 0 \\ -1 & 1 \end{bmatrix} \begin{bmatrix} i_s(t) \\ v_s(t) \end{bmatrix}$$

$$\equiv \dot{\mathbf{X}} = \mathbf{A}\mathbf{X} + \mathbf{B} \begin{bmatrix} i_s(t) \\ v_s(t) \end{bmatrix}, \mathbf{X} = \begin{bmatrix} i_L(t) \\ v_C(t) \end{bmatrix}$$

$$\Rightarrow \mathbf{A} = \begin{bmatrix} -1 & -1 \\ 1 & -1 \end{bmatrix}, \mathbf{B} = \begin{bmatrix} 1 & 0 \\ -1 & 1 \end{bmatrix}$$

Choice (3) is the answer.

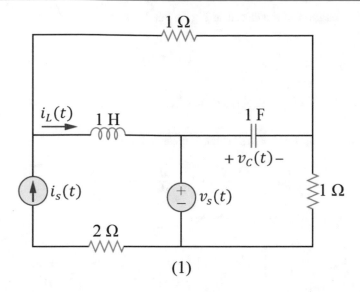

(1)

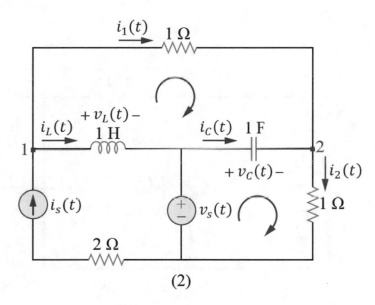

(2)

Figure 2.2 The circuit of solution of problem 2.2

2.3. Recall that the current-voltage relation of inductor and voltage-current relation of capacitor are as follows:

$$v_L(t) = L\frac{d}{dt}i_L(t) \xrightarrow{L=\frac{1}{2}} v_L(t) = \frac{1}{2}\frac{d}{dt}i_L(t) \tag{1}$$

$$i_C(t) = C\frac{d}{dt}v_C(t) \xrightarrow{C=\frac{1}{3}} i_C(t) = \frac{1}{3}\frac{d}{dt}v_C(t) \tag{2}$$

We need to write the state equations of the circuit based on the state vector given in the problem:

$$\mathbf{X} = \begin{bmatrix} i_L(t) \\ v_C(t) \end{bmatrix}$$

Applying KCL in the indicated node of the circuit of Figure 2.3.2:

$$-i_L(t) + 2i(t) + i(t) = 0 \Rightarrow i(t) = \frac{1}{3}i_L(t) \tag{3}$$

Applying KVL in the indicated loop of the circuit of Figure 2.3.2:

$$1 \times i_L(t) + \frac{1}{2}\frac{d}{dt}i_L(t) + 1 \times i(t) + v_C(t) = 0 \overset{(3)}{\Rightarrow} i_L(t) + \frac{1}{2}\frac{d}{dt}i_L(t) + \frac{1}{3}i_L(t) + v_C(t) = 0$$

$$\Rightarrow \frac{d}{dt}i_L(t) = -\frac{8}{3}i_L(t) - 2v_C(t) \qquad (4)$$

Solving (2) and (3) and considering $i_C(t) = i(t)$, which is clear in Figure 2.3.2:

$$\frac{1}{3}i_L(t) = \frac{1}{3}\frac{d}{dt}v_C(t) \Rightarrow \frac{d}{dt}v_C(t) = i_L(t) \qquad (5)$$

By writing (4) and (5) in the form of matrices, we have:

$$\begin{bmatrix} \frac{d}{dt}i_L(t) \\ \frac{d}{dt}v_C(t) \end{bmatrix} = \begin{bmatrix} -\frac{8}{3} & -2 \\ 1 & 0 \end{bmatrix} \begin{bmatrix} i_L(t) \\ v_C(t) \end{bmatrix} \Rightarrow A = \begin{bmatrix} -\frac{8}{3} & -2 \\ 1 & 0 \end{bmatrix}$$

Choice (1) is the answer.

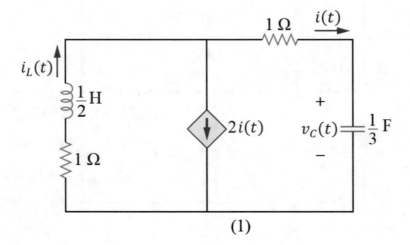

(1)

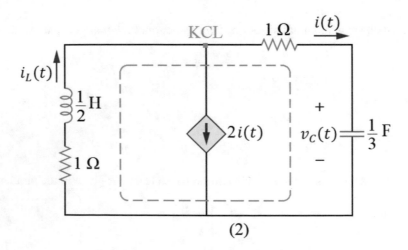

(2)

Figure 2.3 The circuit of solution of problem 2.3

2.4. In this problem, only matrix **A** needs to be determined; therefore, the source can be turned off, as is illustrated in Figure 2.4.2.

Recall that the current-voltage relation of inductor and voltage-current relation of capacitor are as follows:

$$v_L(t) = L\frac{d}{dt}i_L(t) \xrightarrow{L=1} v_L(t) = \frac{d}{dt}i_L(t) \tag{1}$$

$$i_C(t) = C\frac{d}{dt}v_C(t) \xrightarrow{C=1} i_C(t) = \frac{d}{dt}v_C(t) \tag{2}$$

We need to write the state equations of the circuit based on the state vector given in the problem:

$$\mathbf{X} = \begin{bmatrix} v_C(t) \\ i_L(t) \end{bmatrix}$$

Applying KVL in the left-side mesh of the circuit of Figure 2.4.2:

$$1 \times i_C(t) + v_C(t) + 1 \times (i_C(t) - i_L(t)) = 0 \Rightarrow i_C(t) = -0.5v_C(t) + 0.5i_L(t) \tag{3}$$

Solving (2) and (3):

$$2 \times \frac{d}{dt}v_C(t) + v_C(t) - i_L(t) = 0 \Rightarrow \frac{d}{dt}v_C(t) = -0.5v_C(t) + 0.5i_L(t) \tag{4}$$

Applying KVL in the right-side mesh of the circuit of Figure 2.4.2:

$$1 \times (i_L(t) - i_C(t)) + \frac{d}{dt}i_L(t) + 1 \times i_L(t) = 0 \Rightarrow \frac{d}{dt}i_L(t) = -2i_L(t) + i_C(t) \tag{5}$$

Solving (3) and (5):

$$\frac{d}{dt}i_L(t) = -2i_L(t) - 0.5v_C(t) + 0.5i_L(t) = -1.5i_L(t) - 0.5v_C(t) \tag{6}$$

Writing (4) and (6) in the form of matrices:

$$\begin{bmatrix} \frac{d}{dt}v_C(t) \\ \frac{d}{dt}i_L(t) \end{bmatrix} = \begin{bmatrix} -0.5 & 0.5 \\ -0.5 & -1.5 \end{bmatrix} \begin{bmatrix} v_C(t) \\ i_L(t) \end{bmatrix} \Rightarrow \mathbf{A} = \begin{bmatrix} -0.5 & 0.5 \\ -0.5 & -1.5 \end{bmatrix}$$

Choice (3) is the answer.

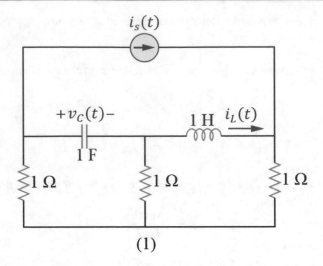

$$(1)$$

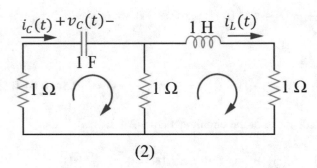

$$(2)$$

Figure 2.4 The circuit of solution of problem 2.4

2.5. Recall that the current-voltage relation of inductor and voltage-current relation of capacitor are as follows:

$$v_L(t) = L\frac{d}{dt}i_L(t) \xrightarrow{L\,=\,2} v_L(t) = 2\frac{d}{dt}i_L(t) \tag{1}$$

$$i_C(t) = C\frac{d}{dt}v_C(t) \xrightarrow{C\,=\,1} i_C(t) = \frac{d}{dt}v_C(t) \tag{2}$$

We need to write the state equations of the circuit based on the state vector given in the problem:

$$\mathbf{X} = \begin{bmatrix} v_C(t) \\ i_L(t) \end{bmatrix}$$

Applying KVL in the indicated loop of the circuit of Figure 2.5.2:

$$-v_s(t) + v_C(t) + 2\frac{d}{dt}i_L(t) = 0 \;\Rightarrow\; \frac{d}{dt}i_L(t) = 0.5v_s(t) - 0.5v_C(t) \tag{3}$$

Applying KCL in the indicated supernode of the circuit of Figure 2.5.2:

$$-\frac{v_C(t)}{2} - i_C(t) + i_L(t) + \frac{2\frac{d}{dt}i_L(t)}{1} = 0 \;\Rightarrow\; -\frac{v_C(t)}{2} - i_C(t) + i_L(t) + 2\frac{d}{dt}i_L(t) = 0$$

$$\overset{(3)}{\Rightarrow} -\frac{v_C(t)}{2} - i_C(t) + i_L(t) + v_s(t) - v_C(t) = 0 \;\Rightarrow\; -1.5v_C(t) - i_C(t) + i_L(t) + v_s(t) = 0$$

$$\overset{(2)}{\Rightarrow} -1.5v_C(t) - \frac{d}{dt}v_C(t) + i_L(t) + v_s(t) = 0 \;\Rightarrow\; \frac{d}{dt}v_C(t) = i_L(t) + v_s(t) - 1.5v_C(t) \tag{4}$$

Writing (3) and (4) in the form of matrices:

$$\begin{bmatrix} \dfrac{d}{dt}v_C(t) \\[2mm] \dfrac{d}{dt}i_L(t) \end{bmatrix} = \begin{bmatrix} -\dfrac{3}{2} & 1 \\[2mm] -\dfrac{1}{2} & 0 \end{bmatrix} \begin{bmatrix} v_C(t) \\[2mm] i_L(t) \end{bmatrix} + \begin{bmatrix} 1 \\[2mm] \dfrac{1}{2} \end{bmatrix} v_s(t)$$

Choice (3) is the answer.

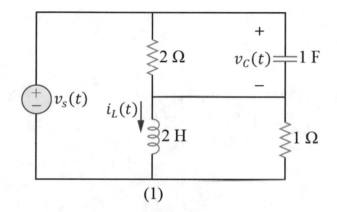

(1)

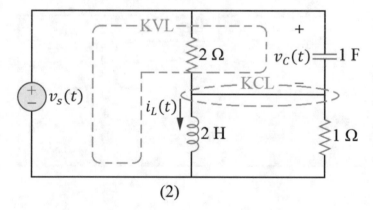

(2)

Figure 2.5   The circuit of solution of problem 2.5

2.6. First, we should simplify the series connection of the inductors as well as the parallel connection of the capacitors, as can be seen in the circuit of Figure 2.6.2.

Recall that the current-voltage relation of inductor and voltage-current relation of capacitor are as follows:

$$v_L(t) = L\frac{d}{dt}i_L(t) \overset{L=1}{\Longrightarrow} v_L(t) = \frac{d}{dt}i_L(t) \tag{1}$$

$$i_C(t) = C\frac{d}{dt}v_C(t) \overset{C=0.5}{\Longrightarrow} i_C(t) = 0.5\frac{d}{dt}v_C(t) \tag{2}$$

We need to write the state equations of the circuit based on the state vector given in the problem:

$$\mathbf{X} = \begin{bmatrix} v_C(t) \\ i_L(t) \end{bmatrix}$$

Applying KCL in the indicated node of the circuit of Figure 2.6.2:

$$-i_C(t) + i_L(t) = 0 \Rightarrow i_C(t) = i_L(t) \tag{3}$$

$$\overset{(2)}{\Rightarrow} 0.5 \frac{d}{dt} v_C(t) = i_L(t) \Rightarrow \frac{d}{dt} v_C(t) = 2i_L(t) \tag{4}$$

Applying KVL in the indicated mesh of the circuit of Figure 2.6.2:

$$-\frac{d}{dt} i_L(t) - 2i_L(t) - v_C(t) + 3 \times (2i_L(t) - i_C(t)) = 0 \Rightarrow -\frac{d}{dt} i_L(t) + 4i_L(t) - v_C(t) - 3i_C(t) = 0$$

$$\overset{(3)}{\Rightarrow} -\frac{d}{dt} i_L(t) + 4i_L(t) - v_C(t) - 3i_L(t) = 0 \Rightarrow \frac{d}{dt} i_L(t) = i_L(t) - v_C(t) \tag{5}$$

Writing (4) and (5) in the form of matrices:

$$\begin{bmatrix} \frac{d}{dt} v_C(t) \\ \frac{d}{dt} i_L(t) \end{bmatrix} = \begin{bmatrix} 0 & 2 \\ -1 & 1 \end{bmatrix} \begin{bmatrix} v_C(t) \\ i_L(t) \end{bmatrix} \Rightarrow A = \begin{bmatrix} 0 & 2 \\ -1 & 1 \end{bmatrix}$$

Choice (1) is the answer.

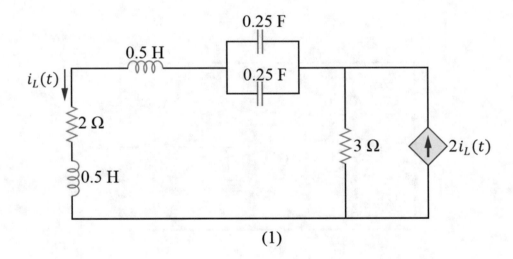

(1)

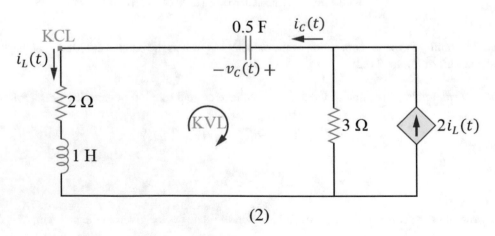

(2)

Figure 2.6 The circuit of solution of problem 2.6

2.7. Recall that the current-voltage relation of inductor and voltage-current relation of capacitor are as follows:

$$v_L(t) = L\frac{d}{dt}i_L(t) \xRightarrow{L=2} v_L(t) = 2\frac{d}{dt}i_L(t)$$

$$i_C(t) = C\frac{d}{dt}v_C(t) \xRightarrow{C=1} i_C(t) = \frac{d}{dt}v_C(t)$$

We need to write the state equations of the circuit based on the state vector given in the problem:

$$\mathbf{X} = \begin{bmatrix} i_L(t) \\ v_C(t) \end{bmatrix}$$

Applying KCL in the indicated supernode of the circuit of Figure 2.7.2:

$$-i_s(t) + \frac{d}{dt}v_C(t) + i_1(t) - \alpha i_L(t) = 0 \Rightarrow i_1(t) - i_s(t) - \frac{d}{dt}v_C(t) + \alpha i_L(t) \tag{1}$$

Applying KVL in the indicated upper mesh of the circuit of Figure 2.7.2:

$$-2\frac{d}{dt}i_L(t) - 1 \times i_2(t) = 0 \Rightarrow i_2(t) = -2\frac{d}{dt}i_L(t) \tag{2}$$

Applying KCL in the indicated node of the circuit of Figure 2.7.2:

$$-i_2(t) + i_L(t) + i_1(t) - \alpha i_L(t) = 0 \Rightarrow i_2(t) = i_1(t) + (1 - \alpha)i_L(t) \tag{3}$$

Applying KVL in the indicated lower mesh of the circuit of Figure 2.7.2:

$$-v_C(t) + 1 \times i_2(t) + 0.5 \times i_1(t) = 0 \Rightarrow i_2(t) = v_C(t) - 0.5i_1(t) \tag{4}$$

Solving (3) and (4):

$$i_1(t) + (1 - \alpha)i_L(t) = v_C(t) - 0.5i_1(t) \Rightarrow 1.5i_1(t) = v_C(t) + (\alpha - 1)i_L(t) = 0$$

$$\Rightarrow i_1(t) - \frac{2}{3}v_C(t) + \frac{2}{3}(\alpha - 1)i_L(t) \tag{5}$$

Solving (4) and (5):

$$i_2(t) = v_C(t) - \frac{1}{3}v_C(t) - \frac{1}{3}(\alpha - 1)i_L(t) \Rightarrow i_2(t) = \frac{2}{3}v_C(t) - \frac{1}{3}(\alpha - 1)i_L(t) \tag{6}$$

Solving (1) and (5):

$$\frac{2}{3}v_C(t) + \frac{2}{3}(\alpha - 1)i_L(t) = i_s(t) - \frac{d}{dt}v_C(t) + \alpha i_L(t) \Rightarrow \frac{d}{dt}v_C(t) = i_s(t) - \frac{2}{3}v_C(t) + \left(\frac{\alpha + 2}{3}\right)i_L(t) \tag{7}$$

Solving (2) and (6):

$$-2\frac{d}{dt}i_L(t) = \frac{2}{3}v_C(t) - \frac{1}{3}(\alpha - 1)i_L(t) \Rightarrow \frac{d}{dt}i_L(t) = -\frac{1}{3}v_C(t) + \frac{1}{6}(\alpha - 1)i_L(t) \tag{8}$$

By writing (7) and (8) in the form of matrices, we have:

$$\begin{bmatrix} \frac{d}{dt}i_L(t) \\ \frac{d}{dt}v_C(t) \end{bmatrix} = \begin{bmatrix} \frac{1}{6}(\alpha - 1) & -\frac{1}{3} \\ \frac{1}{3}(\alpha + 2) & -\frac{2}{3} \end{bmatrix} \begin{bmatrix} i_L(t) \\ v_C(t) \end{bmatrix} + \begin{bmatrix} 0 \\ 1 \end{bmatrix} i_s(t) \Rightarrow \mathbf{B} = \begin{bmatrix} 0 \\ 1 \end{bmatrix}$$

Choice (3) is the answer.

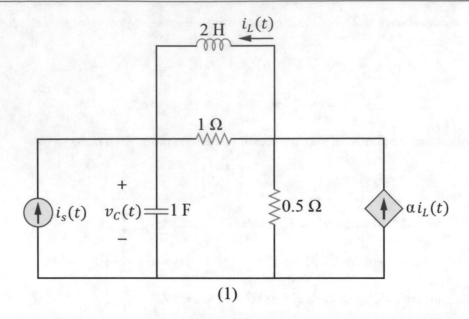

(1)

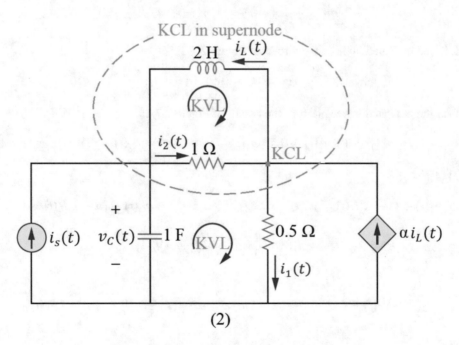

(2)

Figure 2.7 The circuit of solution of problem 2.7

2.8. In this problem, only matrix **A** needs to be determined; therefore, the sources should be turned off, as is illustrated in Figure 2.8.2. As can be seen, the voltage source is replaced by a short circuit branch and the current source is replaced by an open circuit branch.

Recall that the current-voltage relation of inductor and voltage-current relation of capacitor are as follows:

$$v_L(t) = L\frac{d}{dt}i_L(t) \xRightarrow{L=1} v_L(t) = \frac{d}{dt}i_L(t) \tag{1}$$

$$i_C(t) = C\frac{d}{dt}v_C(t) \xRightarrow{C=1} i_C(t) = \frac{d}{dt}v_C(t) \tag{2}$$

We need to write the state equations of the circuit based on the state vector given in the problem:

$$\mathbf{X} = \begin{bmatrix} v_C(t) \\ i_L(t) \end{bmatrix}$$

Applying KVL in the right-side mesh of the circuit of Figure 2.8.2:

$$-\frac{d}{dt}i_L(t) + 1 \times (i_C(t) - i_L(t)) = 0 \Rightarrow -\frac{d}{dt}i_L(t) + i_C(t) - i_L(t) = 0$$

$$\overset{(2)}{\Rightarrow} -\frac{d}{dt}i_L(t) + \frac{d}{dt}v_C(t) - i_L(t) = 0 \tag{3}$$

Applying KVL in the left-side mesh of the circuit of Figure 2.8.2:

$$1 \times i_C(t) + 1 \times i_C(t) + v_C(t) + \frac{d}{dt}i_L(t) = 0 \Rightarrow 2i_C(t) + v_C(t) + \frac{d}{dt}i_L(t) = 0$$

$$\overset{(2)}{\Rightarrow} 2\frac{d}{dt}v_C(t) + v_C(t) + \frac{d}{dt}i_L(t) = 0 \tag{4}$$

Solving (3) and (4):

$$2\frac{d}{dt}v_C(t) + v_C(t) + \frac{d}{dt}v_C(t) - i_L(t) = 0 \Rightarrow \frac{d}{dt}v_C(t) = -\frac{1}{3}v_C(t) + \frac{1}{3}i_L(t) \tag{5}$$

Solving (5) and (4):

$$2\left(-\frac{1}{3}v_C(t) + \frac{1}{3}i_L(t)\right) + v_C(t) + \frac{d}{dt}i_L(t) = 0 \Rightarrow \frac{d}{dt}i_L(t) = -\frac{1}{3}v_C(t) - \frac{2}{3}i_L(t) \tag{6}$$

Writing (5) and (6) in the form of matrices:

$$\begin{bmatrix} \frac{d}{dt}v_C(t) \\ \frac{d}{dt}i_L(t) \end{bmatrix} = \begin{bmatrix} -\frac{1}{3} & \frac{1}{3} \\ -\frac{1}{3} & -\frac{2}{3} \end{bmatrix} \begin{bmatrix} v_C(t) \\ i_L(t) \end{bmatrix} \Rightarrow \mathbf{A} = \begin{bmatrix} -\frac{1}{3} & \frac{1}{3} \\ -\frac{1}{3} & -\frac{2}{3} \end{bmatrix}$$

Choice (3) is the answer.

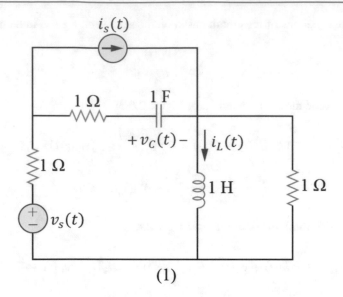

(1)

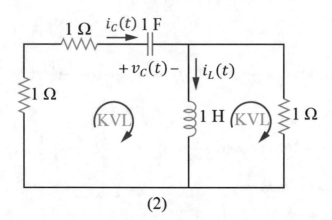

(2)

Figure 2.8 The circuit of solution of problem 2.8

2.9. Since the network includes some resistors, an inductor of 1 H, and a capacitor of 1 F, it can be modeled like the one shown in Figure 2.9. Based on the information given in the problem, we have:

$$\mathbf{X} = \begin{bmatrix} i_L(t) \\ v_C(t) \end{bmatrix} \tag{1}$$

$$\mathbf{A} = \begin{bmatrix} a_{11} & a_{12} \\ a_{21} & a_{22} \end{bmatrix} \tag{2}$$

Therefore, the state equations are as follows:

$$\begin{bmatrix} \dfrac{d}{dt} i_L(t) \\ \dfrac{d}{dt} v_C(t) \end{bmatrix} = \begin{bmatrix} a_{11} & a_{12} \\ a_{21} & a_{22} \end{bmatrix} \begin{bmatrix} i_L(t) \\ v_C(t) \end{bmatrix} \tag{3}$$

As we know, the current-voltage relation of inductor and voltage-current relation of capacitor are as follows:

$$v_L(t) = L\frac{d}{dt} i_L(t) \xrightarrow{L=1} v_L(t) = \frac{d}{dt} i_L(t) \tag{4}$$

$$ i_C(t) = C\frac{d}{dt}v_C(t) \xRightarrow{C=1} i_C(t) = \frac{d}{dt}v_C(t) \tag{5} $$

Solving (3)–(5):

$$ \begin{bmatrix} v_L(t) \\ i_C(t) \end{bmatrix} = \begin{bmatrix} a_{11} & a_{12} \\ a_{21} & a_{22} \end{bmatrix} \begin{bmatrix} i_L(t) \\ v_C(t) \end{bmatrix} \tag{6} $$

Determining the inverse of the matrix:

$$ \begin{bmatrix} i_L(t) \\ v_C(t) \end{bmatrix} = \begin{bmatrix} a_{11} & a_{12} \\ a_{21} & a_{22} \end{bmatrix}^{-1} \begin{bmatrix} v_L(t) \\ i_C(t) \end{bmatrix} = \frac{1}{Det[\mathbf{A}]} \begin{bmatrix} a_{22} & -a_{12} \\ -a_{21} & a_{11} \end{bmatrix} \begin{bmatrix} v_L(t) \\ i_C(t) \end{bmatrix} \tag{7} $$

As can be seen in Figure 2.9.2, by changing the place of the inductor and the capacitor, we have:

$$ i_L(t) = i_C(t) \tag{8} $$

$$ v_L(t) = v_C(t) \tag{9} $$

Solving (7)–(9):

$$ \begin{bmatrix} i_C(t) \\ v_L(t) \end{bmatrix} = \frac{1}{Det[\mathbf{A}]} \begin{bmatrix} a_{22} & -a_{12} \\ -a_{21} & a_{11} \end{bmatrix} \begin{bmatrix} v_C(t) \\ i_L(t) \end{bmatrix} \tag{10} $$

Solving (4), (5), and (10):

$$ \begin{bmatrix} \dfrac{d}{dt}v_C(t) \\ \dfrac{d}{dt}i_L(t) \end{bmatrix} = \frac{1}{Det[\mathbf{A}]} \begin{bmatrix} a_{22} & -a_{12} \\ -a_{21} & a_{11} \end{bmatrix} \begin{bmatrix} v_C(t) \\ i_L(t) \end{bmatrix} \tag{11} $$

Equation (11) shows the state equations of a circuit with the state vector of $\begin{bmatrix} v_C(t) \\ i_L(t) \end{bmatrix}$. Therefore:

$$ \mathbf{A}_{new} = \frac{1}{Det[\mathbf{A}]} \begin{bmatrix} a_{22} & -a_{12} \\ -a_{21} & a_{11} \end{bmatrix} $$

Choice (2) is the answer.

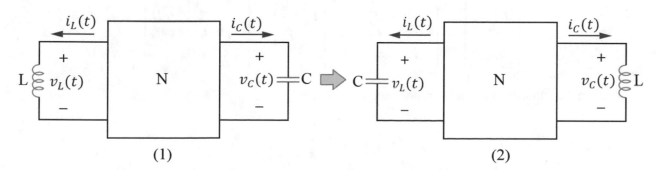

Figure 2.9 The circuit of solution of problem 2.9

2.10. The current-voltage relation of inductor and voltage-current relation of capacitor are as follows:

$$v_L(t) = L\frac{d}{dt}i_L(t)$$

$$i_C(t) = C\frac{d}{dt}v_C(t)$$

We need to write the state equations of the circuit based on the state vector and the input vector given in the problem:

$$\mathbf{X} = \begin{bmatrix} i_{L1}(t) \\ i_{L2}(t) \\ v_C(t) \end{bmatrix}, \mathbf{W} = \begin{bmatrix} i_s(t) \\ v_s(t) \end{bmatrix}$$

Applying KCL in the supernode of the circuit of Figure 2.10.2:

$$-i_R(t) + i_{L1}(t) - i_s(t) + i_{L2}(t) = 0 \Rightarrow i_R(t) = i_{L1}(t) - i_s(t) + i_{L2}(t) \tag{1}$$

Applying KVL in the indicated left-side mesh of the circuit of Figure 2.10.2:

$$-v_s(t) + i_R(t)R + L_1\frac{d}{dt}i_{L1}(t) = 0$$

$$\stackrel{(1)}{\Rightarrow}\frac{d}{dt}i_{L1}(t) = \frac{1}{L_1}v_s(t) - \frac{R}{L_1}i_{L1}(t) + \frac{R}{L_1}i_s(t) - \frac{R}{L_1}i_{L2}(t) \tag{2}$$

Applying KVL in the loop of the circuit of Figure 2.10.2:

$$-v_s(t) + i_R(t)R + v_C(t) + L_2\frac{d}{dt}i_{L2}(t) = 0$$

$$\stackrel{(1)}{\Rightarrow}\frac{d}{dt}i_{L2}(t) = \frac{1}{L_2}v_s(t) - \frac{R}{L_2}i_{L1}(t) + \frac{R}{L_2}i_s(t) - \frac{R}{L_2}i_{L2}(t) - \frac{1}{L_2}v_C(t) \tag{3}$$

Applying KCL in the node of the circuit of Figure 2.10.2:

$$-C\frac{d}{dt}v_C(t) - i_s(t) + i_{L2}(t) = 0 \Rightarrow \frac{d}{dt}v_C(t) = -\frac{1}{C}i_s(t) + \frac{1}{C}i_{L2}(t) \tag{4}$$

Writing (2)–(4) in the form of matrices:

$$\begin{bmatrix} \dfrac{d}{dt}i_{L1}(t) \\ \dfrac{d}{dt}i_{L2}(t) \\ \dfrac{d}{dt}v_C(t) \end{bmatrix} = \begin{bmatrix} -\dfrac{R}{L_1} & -\dfrac{R}{L_1} & 0 \\ -\dfrac{R}{L_2} & -\dfrac{R}{L_2} & -\dfrac{1}{L_2} \\ 0 & \dfrac{1}{C} & 0 \end{bmatrix}\begin{bmatrix} i_{L1}(t) \\ i_{L2}(t) \\ v_C(t) \end{bmatrix} + \begin{bmatrix} \dfrac{R}{L_1} & \dfrac{1}{L_1} \\ \dfrac{R}{L_2} & \dfrac{1}{L_2} \\ -\dfrac{1}{C} & 0 \end{bmatrix}\begin{bmatrix} i_s(t) \\ v_s(t) \end{bmatrix} \tag{6}$$

Based on the information given in the problem, we know that:

$$\mathbf{A} = \begin{bmatrix} -1 & -1 & 0 \\ -\dfrac{1}{2} & -\dfrac{1}{2} & -\dfrac{1}{2} \\ 0 & 1 & 0 \end{bmatrix} \tag{7}$$

By comparing (6) and (7), we can write:

$$\begin{cases} -\dfrac{R}{L_1} = -1 \\[2mm] -\dfrac{R}{L_2} = -\dfrac{1}{2} \\[2mm] -\dfrac{1}{L_2} = -\dfrac{1}{2} \\[2mm] \dfrac{1}{C} = 1 \end{cases} \Rightarrow R = 1\,\Omega, L_1 = 1\,H, L_2 = 2\,H, C = 1\,F \tag{8}$$

Solving (6) and (8), we can write:

$$\mathbf{B} = \begin{bmatrix} \dfrac{1}{1} & \dfrac{1}{1} \\[2mm] \dfrac{1}{2} & \dfrac{1}{2} \\[2mm] -\dfrac{1}{1} & 0 \end{bmatrix} \Rightarrow \mathbf{B} = \begin{bmatrix} 1 & 1 \\[1mm] \dfrac{1}{2} & \dfrac{1}{2} \\[1mm] -1 & 0 \end{bmatrix}$$

Choice (4) is the answer.

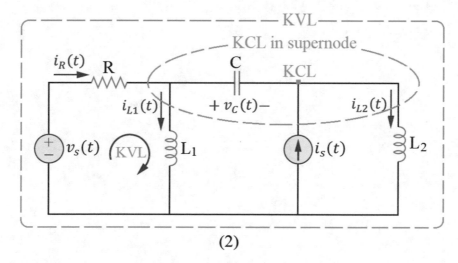

(1)

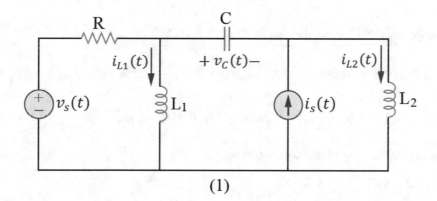

(2)

Figure 2.10   The circuit of solution of problem 2.10

2.11. The current-voltage relation of inductor and voltage-current relation of capacitor are as follows:

$$v_L(t) = L\frac{d}{dt}i_L(t) \xRightarrow{L\,=\,1} v_L(t) = \frac{d}{dt}i_L(t)$$

$$i_C(t) = C\frac{d}{dt}v_C(t) \xRightarrow{C\,=\,1} i_C(t) = \frac{d}{dt}v_C(t)$$

We need to write the state equations of the circuit based on the state vector given in the problem:

$$\mathbf{X} = \begin{bmatrix} v_C(t) \\ i_L(t) \end{bmatrix}$$

Applying KVL in the indicated mesh of the circuit of Figure 2.11.2:

$$-v_C(t) + 1 \times i_R(t) + 1 \times (i_R(t) - i_L(t)) = 0 \Rightarrow -v_C(t) + 2i_R(t) - i_L(t) = 0 \Rightarrow i_R(t) = \frac{1}{2}v_C(t) + \frac{1}{2}i_L(t) \qquad (1)$$

Applying KCL in the indicated node of the circuit of Figure 2.11.2:

$$-i_s(t) + i_R(t) + \frac{d}{dt}v_C(t) = 0 \overset{(1)}{\Rightarrow} \frac{d}{dt}v_C(t) = i_s(t) - \frac{1}{2}v_C(t) - \frac{1}{2}i_L(t) \qquad (2)$$

Applying KVL in the indicated loop of the circuit of Figure 2.11.2:

$$-v_C(t) + 1 \times i_R(t) + \frac{d}{dt}i_L(t) + 1 \times (i_L(t) - i_s(t)) = 0 \Rightarrow \frac{d}{dt}i_L(t) = v_C(t) - i_R(t) - i_L(t) + i_s(t)$$

$$\overset{(1)}{\Rightarrow} \frac{d}{dt}i_L(t) = v_C(t) - \frac{1}{2}v_C(t) - \frac{1}{2}i_L(t) - i_L(t) + i_s(t) = \frac{1}{2}v_C(t) - \frac{3}{2}i_L(t) + i_s(t) \qquad (3)$$

By writing (2) and (3) in the form of matrices, we have:

$$\begin{bmatrix} \dfrac{d}{dt}v_C(t) \\[2mm] \dfrac{d}{dt}i_L(t) \end{bmatrix} = \begin{bmatrix} -\dfrac{1}{2} & -\dfrac{1}{2} \\[2mm] \dfrac{1}{2} & -\dfrac{3}{2} \end{bmatrix}\begin{bmatrix} v_C(t) \\[2mm] i_L(t) \end{bmatrix} + \begin{bmatrix} 1 \\[2mm] 1 \end{bmatrix}i_s(t) \Rightarrow \mathbf{A} = \begin{bmatrix} -0.5 & -0.5 \\ 0.5 & -1.5 \end{bmatrix}$$

Choice (3) is the answer.

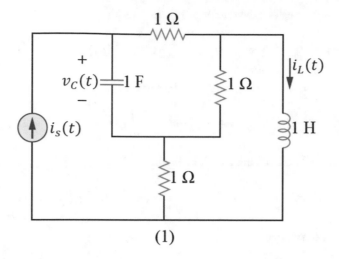

(1)

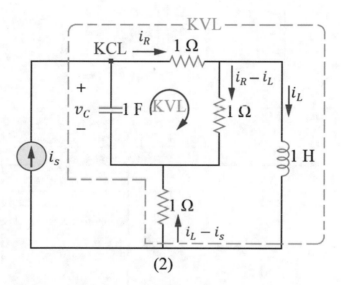

(2)

Figure 2.11   The circuit of solution of problem 2.11

2.12. First, we should use source transformation theorem for the parallel connection of the independent current source and the 2-Ω resistor, as is shown in Figure 2.12.2.

Recall that the current-voltage relation of inductor and voltage-current relation of capacitor are as follows:

$$v_L(t) = L\frac{d}{dt}i_L(t)$$

$$i_C(t) = C\frac{d}{dt}v_C(t)$$

We need to write the state equations of the circuit based on the state vector given in the problem:

$$\mathbf{X} = \begin{bmatrix} i_1(t) \\ i_2(t) \\ v(t) \end{bmatrix}$$

Applying KVL in the left-side mesh of the circuit of Figure 2.12.2:

$$-2i_s(t) + 2 \times i_1(t) + 0.5 \times \frac{d}{dt}i_1(t) + v(t) = 0 \implies \frac{d}{dt}i_1(t) = 4i_s(t) - 4i_1(t) - 2v(t) \tag{1}$$

Applying KVL in the right-side mesh of the circuit of Figure 2.12.2:

$$-v(t) + 0.5 \times \frac{d}{dt}i_2(t) + 2i_2(t) + v_s(t) = 0 \implies \frac{d}{dt}i_2(t) = -4i_2(t) - 2v_s(t) + 2v(t) \tag{2}$$

Applying KCL in the indicated node of the circuit of Figure 2.12.2:

$$-i_1(t) + i_2(t) + 1 \times \frac{d}{dt}v(t) = 0 \implies \frac{d}{dt}v(t) = i_1(t) - i_2(t) \tag{3}$$

Writing (1)–(3) in the form of matrices:

$$\begin{bmatrix} \frac{d}{dt}i_1(t) \\ \frac{d}{dt}i_2(t) \\ \frac{d}{dt}v(t) \end{bmatrix} = \begin{bmatrix} -4 & 0 & -2 \\ 0 & -4 & 2 \\ 1 & -1 & 0 \end{bmatrix} \begin{bmatrix} i_1(t) \\ i_2(t) \\ v(t) \end{bmatrix} + \begin{bmatrix} 4 & 0 \\ 0 & -2 \\ 0 & 0 \end{bmatrix} \begin{bmatrix} i_s(t) \\ v_s(t) \end{bmatrix}$$

Choice (3) is the answer.

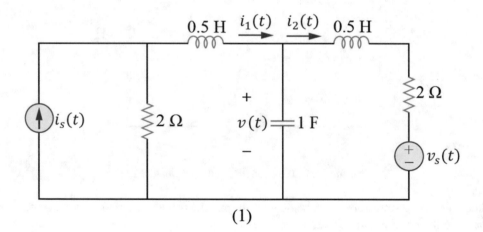

(1)

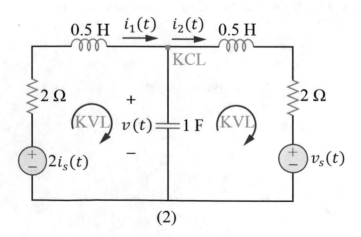

(2)

Figure 2.12 The circuit of solution of problem 2.12

References

1. Rahmani-Andebili, M. (2020). DC Electrical circuit analysis: Practice problems, methods, and solutions, Springer Nature.
2. Rahmani-Andebili, M. (2020). AC Electrical circuit analysis: Practice problems, methods, and solutions, Springer Nature.

Problems: Laplace Transform and Network Function 3

Abstract

In this chapter, Laplace transform and network function (transfer function) are applied to solve the basic and advanced problems of electrical circuit analysis. In this chapter, the problems are categorized in different levels based on their difficulty levels (easy, normal, and hard) and calculation amounts (small, normal, and large). Additionally, the problems are ordered from the easiest problem with the smallest computations to the most difficult problems with the largest calculations.

3.1. Determine the network function (transfer function) of the circuit illustrated in Figure 3.1 [1–2].

Difficulty level ● Easy ○ Normal ○ Hard
Calculation amount ● Small ○ Normal ○ Large

1. $\frac{V_{out}(s)}{V_{in}(s)} = \frac{1}{1+RCs}$

2. $\frac{V_{out}(s)}{V_{in}(s)} = \frac{RCs}{1+RCs}$

3. $\frac{V_{out}(s)}{V_{in}(s)} = \frac{1+2RCs}{2+RCs}$

4. $\frac{V_{out}(s)}{V_{in}(s)} = \frac{2RCs}{1+RCs}$

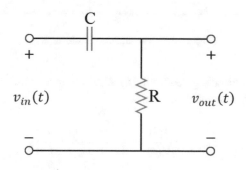

Figure 3.1 The circuit of problem 3.1

3.2. In the circuit of Figure 3.2, determine the input impedance in Laplace domain seen from the left side of the circuit.

Difficulty level ● Easy ○ Normal ○ Hard
Calculation amount ● Small ○ Normal ○ Large

1. $\frac{s+2}{5s+1}\ \Omega$

2. $\frac{s+1}{s+2}\ \Omega$

3. $\frac{s}{5s+2}\ \Omega$

4. $\frac{s+1}{5s+1}\ \Omega$

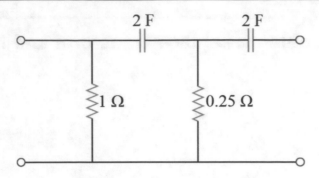

Figure 3.2 The circuit of problem 3.2

3.3. In the circuit of Figure 3.3, determine the network function (transfer function) of $H(s) = \frac{V_{out}(s)}{V_{in}(s)}$:

Difficulty level   ● Easy   ○ Normal   ○ Hard
Calculation amount   ● Small   ○ Normal   ○ Large

1. $\frac{1}{s^2+1}$

2. $\frac{1}{s-1}$

3. $\frac{1}{s^2-1}$

4. $\frac{1}{s+1}$

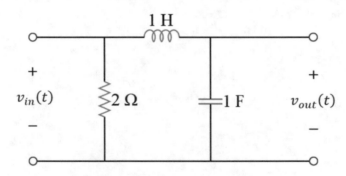

Figure 3.3 The circuit of problem 3.3

3.4. In an electrical circuit, the relation below is given for the input admittance of the circuit in Laplace domain. Determine the differential equation between the voltage and current.

$$Y_{in}(s) = \frac{I(s)}{V(s)} = \frac{s^2 + 2s + 3}{4s^3 + 5s^2 + 6s + 7}$$

Difficulty level   ● Easy   ○ Normal   ○ Hard
Calculation amount   ● Small   ○ Normal   ○ Large

1. $4\frac{d^3}{dt^3}i(t) + \frac{d^2}{dt^2}i(t) + 6\frac{d}{dt}i(t) = \frac{d^2}{dt^2}v(t) + \frac{d}{dt}v(t) + 3v(t)$

2. $\frac{d^3}{dt^3}i(t) + \frac{d^2}{dt^2}i(t) + \frac{d}{dt}i(t) = \frac{d^2}{dt^2}v(t) + 2\frac{d}{dt}v(t) + v(t)$

3. $4\frac{d^3}{dt^3}i(t) + 5\frac{d^2}{dt^2}i(t) + 6\frac{d}{dt}i(t) + 7i(t) = \frac{d^2}{dt^2}v(t) + 2\frac{d}{dt}v(t) + 3v(t)$

4. $4\frac{d^3}{dt^3}i(t) + 5\frac{d^2}{dt^2}i(t) + 6\frac{d}{dt}i(t) + 7i(t) = \frac{d^2}{dt^2}v(t) + 2\frac{d}{dt}v(t) + v(t)$

3.5. In the circuit of Figure 3.4, determine the value of $\frac{V(s)}{I(s)}$.

Difficulty level   ● Easy   ○ Normal   ○ Hard
Calculation amount   ● Small   ○ Normal   ○ Large

1. $\frac{s^3-10s^2+10s-1}{s^2+6s+1}$ Ω

2. $\frac{s^3+10s^2+10s+2}{s^2+6s+1}$ Ω

3. $\frac{s^3+10s^2+1}{s^2+3}$ Ω

4. $\frac{s^3-8s^2+8s+2}{s^2+6s+3}$ Ω

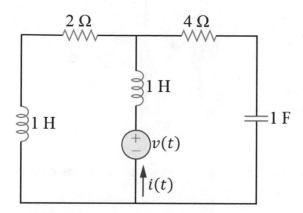

Figure 3.4 The circuit of problem 3.5

3.6. In the circuit of Figure 3.5, determine the network function (transfer function) of $H(s) = \frac{I_L(s)}{I(s)}$.

Difficulty level   ● Easy   ○ Normal   ○ Hard
Calculation amount   ● Small   ○ Normal   ○ Large

1. $\frac{2}{s^2+2s+1}$

2. $\frac{2}{s^2+3s+2}$

3. $\frac{1}{s^2+s+2}$

4. $\frac{1}{s^2+3s+1}$

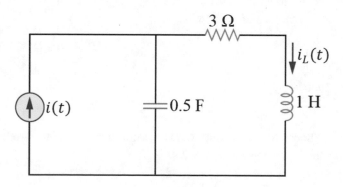

Figure 3.5 The circuit of problem 3.6

3.7. In the circuit of Figure 3.6, determine the input impedance in Laplace domain seen by the terminal ($Z_{in}(s)$).

Difficulty level   ○ Easy   ● Normal   ○ Hard
Calculation amount   ● Small   ○ Normal   ○ Large

1. $2(s^2 + 10s + 5)$ Ω

2. $\frac{1}{5(s+1)}$ Ω

3. $2(s + 1)$ Ω

4. $\frac{2(s^2+10s+5)}{5(s+1)}$ Ω

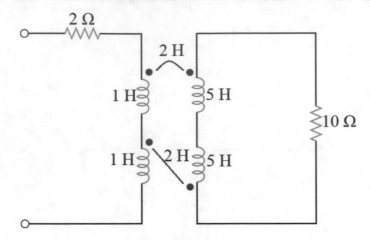

Figure 3.6 The circuit of problem 3.7

3.8. Which one of the choices is correct for the unit step response of a circuit with the following network function (transfer function)?

$$H(s) = \frac{V_{out}(s)}{V_{in}(s)} = \frac{25}{s^2 + 10s + 125}$$

Difficulty level   ○ Easy   ● Normal   ○ Hard
Calculation amount   ● Small   ○ Normal   ○ Large

1. $v_{out}(t = 0^+) = v_{out}(t = \infty) = 0$
2. $v_{out}(t = 0^+) = v_{out}(t = \infty) = 0.2$
3. $v_{out}(t = 0^+) = 0, v_{out}(t = \infty) = 0.2$
4. None of them

3.9. The impulse function of a linear time-invariant (LTI) system is $h(t) = (e^{-t} - e^{-2t})u(t)$. Determine the output response of the system for the input signal of $x(t) = 2e^{-3t}u(t)$.

Difficulty level   ○ Easy   ● Normal   ○ Hard
Calculation amount   ● Small   ○ Normal   ○ Large

1. $y(t) = (e^{-t} + e^{-2t} + e^{-3t})u(t)$
2. $y(t) = (e^{-t} - 2e^{-2t} + e^{-3t})u(t)$
3. $y(t) = (2e^{-t} + e^{-2t} + e^{-3t})u(t)$
4. $y(t) = (e^{-t} + 2e^{-2t} + e^{-3t})u(t)$

3.10. The impulse function of a linear time-invariant (LTI) system is $h(t) = \frac{3}{4}(e^{-t} + e^{-3t})u(t)$. Determine the output response of the system for the input signal of $x(t) = 2\delta(t - 5)$.

Difficulty level   ○ Easy   ● Normal   ○ Hard
Calculation amount   ● Small   ○ Normal   ○ Large

1. $y(t) = (e^{-(t-5)} + e^{-3(t-5)})u(t - 5)$
2. $y(t) = \frac{3}{2}(e^{-(t-5)} + e^{-3(t-5)})u(t - 5)$
3. $y(t) = -\frac{3}{2}(e^{-(t-5)} + e^{-3(t-5)})u(t - 5)$
4. $y(t) = -(e^{-(t-5)} + e^{-3(t-5)})u(t - 5)$

3.11. In the circuit of Figure 3.7, calculate the network function (transfer function) of $\frac{V_2(s)}{V_s(s)}$.

Difficulty level   ○ Easy   ● Normal   ○ Hard
Calculation amount   ○ Small   ● Normal   ○ Large

1. $\dfrac{2(s^2+1)}{(s+1)^2}$

2. $\dfrac{(s^2+1)}{(s+2)^2}$

3. $\dfrac{(s^2+1)}{s^2+s+1}$

4. $\dfrac{(s^2+1)}{(s+1)^2}$

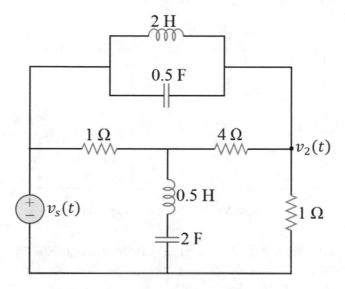

Figure 3.7 The circuit of problem 3.11

3.12. In the circuit of Figure 3.8, $i_s(t)$ is the input and $i(t)$ is the response of the circuit. Determine the impulse response of the circuit.

Difficulty level　　○ Easy　　● Normal　　○ Hard
Calculation amount　○ Small　　● Normal　　○ Large

1. $\left(\frac{1}{3}e^{-5t} - \frac{4}{3}e^{-2t}\right)u(t) - \delta(t)$
2. $\left(\frac{1}{3}e^{-5t} + \frac{4}{3}e^{-2t}\right)u(t) + \delta(t)$
3. $\left(\frac{1}{3}e^{-5t} - \frac{4}{3}e^{-2t}\right)u(t) + \delta(t)$
4. $\left(\frac{1}{3}e^{-5t} + \frac{4}{3}e^{-2t}\right)u(t) - \delta(t)$

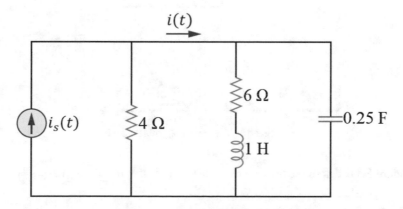

Figure 3.8 The circuit of problem 3.12

3.13. In the circuit of Figure 3.9, determine the differential equation between $i_s(t)$ and $v_a(t)$.

1. $\frac{d^2}{dt^2}v_a(t) + 2\frac{d}{dt}v_a(t) + 2v_a(t) = \frac{d^2}{dt^2}i_s(t) + \frac{d}{dt}i_s(t) + i_s(t)$

2. $\frac{d^2}{dt^2}v_a(t) + \frac{d}{dt}v_a(t) + 2v_a(t) = \frac{d^2}{dt^2}i_s(t) + \frac{d}{dt}i_s(t) + i_s(t)$

3. $\frac{d^2}{dt^2}v_a(t) + \frac{d}{dt}v_a(t) + v_a(t) = \frac{d^2}{dt^2}i_s(t) + \frac{d}{dt}i_s(t) + i_s(t)$

4. $2\frac{d^2}{dt^2}v_a(t) + \frac{d}{dt}v_a(t) + 2v_a(t) = \frac{d^2}{dt^2}i_s(t) + \frac{d}{dt}i_s(t) + i_s(t)$

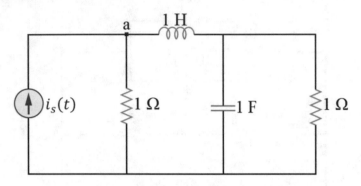

Figure 3.9 The circuit of problem 3.13

3.14. In the circuit of Figure 3.10, determine the differential equation between $v_s(t)$ and $i(t)$.

1. $\frac{d^2}{dt^2}i(t) + \frac{d}{dt}i(t) + i(t) = \frac{d}{dt}v_s(t) + v_s(t)$

2. $\frac{d^2}{dt^2}i(t) + \frac{d}{dt}i(t) + 2i(t) = \frac{d}{dt}v_s(t) + v_s(t)$

3. $\frac{d^2}{dt^2}i(t) + 2\frac{d}{dt}i(t) + 2i(t) = \frac{d}{dt}v_s(t) + v_s(t)$

4. $\frac{d^2}{dt^2}i(t) + 2\frac{d}{dt}i(t) + 2i(t) = 2\frac{d}{dt}v_s(t) + v_s(t)$

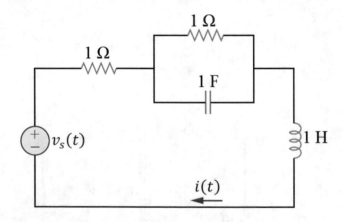

Figure 3.10 The circuit of problem 3.14

3.15. In the circuit of Figure 3.11, determine the network function (transfer function) of $H(s) = \frac{V_{out}(s)}{V_{in}(s)}$.

1. $\frac{s^2+3s}{(s+1)^2}$

2. $\frac{(s-1)^2}{(s+1)^2}$

3. $\frac{s(s-1)}{(s+1)^2}$

4. $\frac{s+3}{(s+1)^2}$

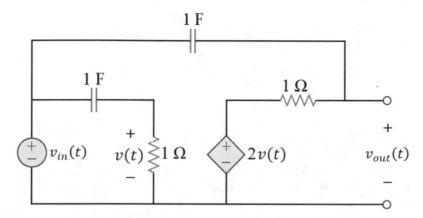

Figure 3.11 The circuit of problem 3.15

3.16. In the circuit of Figure 3.12, determine the equivalent impedance in Laplace domain seen by the terminal ($Z_{eq}(s)$).

Difficulty level   ○ Easy   ● Normal   ○ Hard
Calculation amount   ○ Small   ● Normal   ○ Large

1. $\frac{s-1}{5}$ Ω
2. $\frac{s+11}{5}$ Ω
3. $(s-1)$ Ω
4. 1 Ω

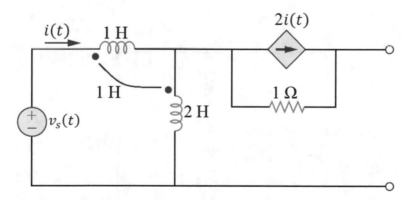

Figure 3.12 The circuit of problem 3.16

3.17. In the circuit of Figure 3.13, determine the network function (transfer function) of $H(s) = \frac{V_{out}(s)}{V_{in}(s)}$.

Difficulty level   ○ Easy   ● Normal   ○ Hard
Calculation amount   ○ Small   ● Normal   ○ Large

1. $\frac{2s+1}{12s+1}$
2. $\frac{2s+1}{12s+4}$
3. $\frac{4s+1}{12s+4}$
4. $\frac{4s+1}{12s+1}$

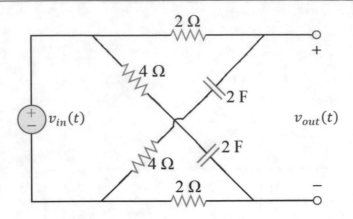

Figure 3.13 The circuit of problem 3.17

3.18. The impulse function of a linear time-invariant (LTI) system is $h(t) = (e^{-t} - e^{-2t})u(t)$. Determine the output response of the system if the input signal is $x(t) = 2e^{-2t}u(t)$.

Difficulty level   ○ Easy   ● Normal   ○ Hard
Calculation amount   ○ Small   ● Normal   ○ Large

1. $y(t) = (2e^{-t} - e^{-2t} + e^{-3t})u(t)$
2. $y(t) = (2e^{-t} - 4e^{-2t} + e^{-3t})u(t)$
3. $y(t) = (2e^{-t} - 4e^{-2t} + 2e^{-3t})u(t)$
4. $y(t) = (2e^{-t} - 4e^{-2t} - 2e^{-3t})u(t)$

3.19. In the circuit of Figure 3.14, calculate the value of $I_L(s)$ for $v_C(0^-) = 2\ V$, $i_L(0^-) = 1\ A$.

Difficulty level   ○ Easy   ● Normal   ○ Hard
Calculation amount   ○ Small   ● Normal   ○ Large

1. $\frac{s^2+2s-20}{s(s^2+2s+2)}\ A$
2. $\frac{s^2+2s+20}{s(s^2+2s+2)}\ A$
3. $\frac{s^2+3s-20}{s(s^2+2s-2)}\ A$
4. $\frac{s^2-3s-20}{s(s^2+2s+2)}\ A$

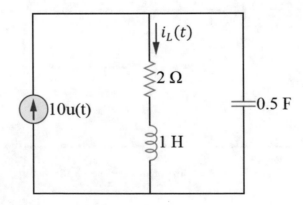

Figure 3.14 The circuit of problem 3.19

3.20. In the circuit of Figure 3.15, calculate the value of $V_{out}(s)$.

Difficulty level   ○ Easy   ● Normal   ○ Hard
Calculation amount   ○ Small   ● Normal   ○ Large

1. $\frac{5s}{s^2(s^3+s^2+3s+1)}$ V

2. $\frac{5s^2}{s^2(s^3+2s^2+3s+1)}$ V

3. $\frac{5s^2}{(s^3+s^2+s+1)}$ V

4. $\frac{5s}{s^2(s^3+2s^2+3s+1)}$ V

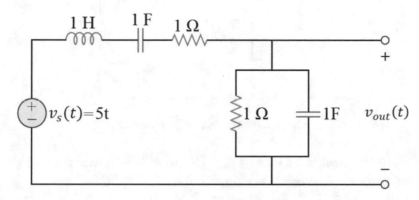

Figure 3.15 The circuit of problem 3.20

3.21. In the circuit of Figure 3.16, calculate the impulse response of $v_{out}(t)$.

Difficulty level   ○ Easy   ● Normal   ○ Hard
Calculation amount   ○ Small   ○ Normal   ● Large

1. $(2e^{-t} - e^{-0.5t})u(t)$ V
2. $2te^{-t}u(t)$ V
3. $(e^{-t} - e^{-0.5t})u(t)$ V
4. $2e^{-t}u(t)$ V

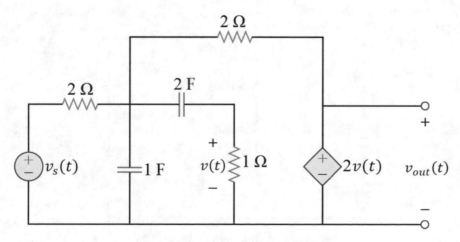

Figure 3.16 The circuit of problem 3.21

3.22. In the circuit of Figure 3.17, determine the network function (transfer function) of $\frac{V_{out}(s)}{V_{in}(s)}$.

Difficulty level   ○ Easy   ● Normal   ○ Hard
Calculation amount   ○ Small   ○ Normal   ● Large

1. $\frac{1}{s^4+s^3+s^2+3s+1}$

2. $\frac{1}{s^4+2s^3+4s^2+3s+1}$

3. $\frac{1}{s^3+s^2+3s+1}$

4. $\frac{1}{s^4-2s^3-3s^2+3s+1}$

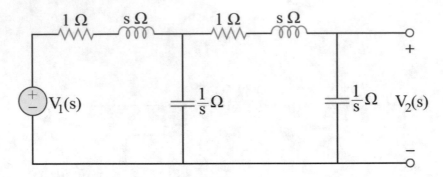

Figure 3.17 The circuit of problem 3.22

3.23. In the circuit of Figure 3.18, calculate the value of $V_{out}(s)$ if $v_s(t) = A\cos(\omega t)$ and the circuit is at zero state.

Difficulty level   ○ Easy   ● Normal   ○ Hard
Calculation amount   ○ Small   ○ Normal   ● Large

1. $\dfrac{A\left(\frac{1}{R_2}-R_1\right)s^2}{\left(s^2+\frac{R_1}{L}s+R_2\right)(s^2+\omega^2)}\,V$

2. $\dfrac{A\left(\frac{1}{R_2C}-\frac{R_1}{L}\right)s^2}{\left(s^2+\left(\frac{R_1}{L}+\frac{1}{R_2C}\right)s+\frac{R_1}{LR_2C}\right)(s^2+\omega^2)}\,V$

3. $\dfrac{A\left(\frac{1}{R_2}-R_1\right)s^2}{\left(s^2+\frac{R_1}{L}s+R_2\right)(s^2-\omega^2)}\,V$

4. $\dfrac{\left(\frac{1}{R_2}+R_1\right)s^2}{\left(s^2+\frac{R_1}{L}s+R_2\right)(s^2+\omega^2)}\,V$

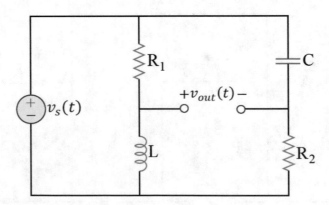

Figure 3.18 The circuit of problem 3.23

3.24. In the circuit of Figure 3.19, the input impedance in Laplace domain is as follows:

$$Z_{in}(s) = \frac{s^2+s+2}{2s^2+s+1}$$

While the circuit is in zero state, the switch is closed at $t=0$, and then $i(0^+)=6\,A$ is measured. Determine the value of E.

Difficulty level   ○ Easy   ○ Normal   ● Hard
Calculation amount   ● Small   ○ Normal   ○ Large

1. 3 V
2. 6 V
3. 9 V
4. 12 V

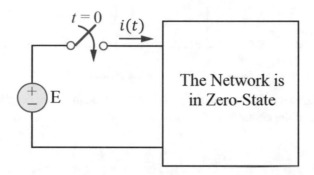

Figure 3.19 The circuit of problem 3.24

3.25. In the circuit of Figure 3.20, $v_s(t)$ is a DC voltage source. Determine the time (in second) that the voltage of the capacitor will be twice as the voltage of the source. The primary voltage of the capacitor and the primary current of the inductor are zero.

Difficulty level   ○ Easy   ○ Normal   ● Hard
Calculation amount   ○ Small   ● Normal   ○ Large

1. $\pi\sqrt{LC}$ sec.
2. $2\pi\sqrt{LC}$ sec.
3. $\frac{\sqrt{LC}}{\pi}$ sec.
4. No time can be found, as it is impossible.

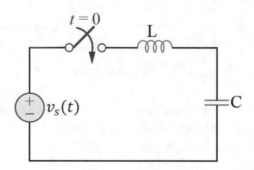

Figure 3.20 The circuit of problem 3.25

3.26. In the circuit of Figure 3.21, determine the Thevenin equivalent circuit in Laplace domain seen from the terminal.

Difficulty level   ○ Easy   ○ Normal   ● Hard
Calculation amount   ○ Small   ● Normal   ○ Large

1. $Z_{Th}(s) = \frac{3}{s+1}$ Ω, $V_{Th}(s) = 3$ V
2. $Z_{Th}(s) = (3s + 3)$ Ω, $V_{Th}(s) = \frac{3}{s}$ V
3. $Z_{Th}(s) = \frac{3s}{s+1}$ Ω, $V_{Th}(s) = \frac{3}{s}$ V
4. $Z_{Th}(s) = (3s + 3)$ Ω, $V_{Th}(s) = 3$ V

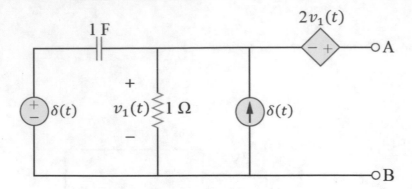

Figure 3.21 The circuit of problem 3.26

3.27. In the circuit of Figure 3.22, determine the energy stored in the inductor at $t = \infty$ if $i_1(t) = \sqrt{2}u(t)$, and the network function (transfer function) of the circuit is as follows:

$$H(s) = \frac{I_2(s)}{I_1(s)} = \frac{4(s+20)}{s+8}$$

Difficulty level   ○ Easy   ○ Normal   ● Hard
Calculation amount   ○ Small   ● Normal   ○ Large

1. 0.1 J
2. 0.15 J
3. 0.2 J
4. 0.05 J

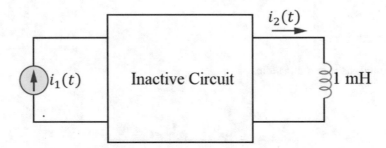

Figure 3.22 The circuit of problem 3.27

3.28. In the circuit of Figure 3.23, determine the capacitance (C) and the primary voltage of the capacitor ($v_C(0^-)$) to have $i_{out}(t > 0) = 0$. The inductor does not have any primary energy and $v_s(t) = \sin(2t)u(t)$.

Difficulty level   ○ Easy   ○ Normal   ● Hard
Calculation amount   ○ Small   ● Normal   ○ Large

1. $C = \frac{1}{4} F, v_C(0^-) = -8 V$
2. $C = \frac{1}{8} F, v_C(0^-) = -4 V$
3. $C = \frac{1}{8} F, v_C(0^-) = 8 V$
4. $C = \frac{1}{4} F, v_C(0^-) = 8 V$

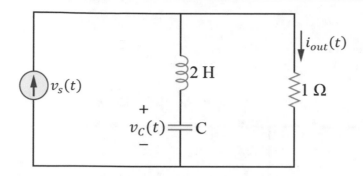

Figure 3.23 The circuit of problem 3.28

3.29. In an electrical circuit, the relation below exists. If $v_{in}(t) = 4 \cos (2t)$, determine the steady-state output voltage of the circuit:

$$H(s) = \frac{V_{out}(s)}{V_{in}(s)} = \frac{10(s+1)}{s^2 + 2s + 3}$$

Difficulty level   ○ Easy   ○ Normal   ● Hard
Calculation amount   ○ Small   ● Normal   ○ Large

1. $v_{out}(t) = 21.76 \cos (2t + 40.6°) \, V$
2. $v_{out}(t) = 20.2 \cos (2t + 40.6°) \, V$
3. $v_{out}(t) = 21.76 \cos (2t - 40.6°) \, V$
4. $v_{out}(t) = 43.52 \cos (2t - 40.6°) \, V$

3.30. The pulse voltage, shown in the circuit of Figure 3.24.1, is applied on the circuit of Figure 3.24.2. Determine the current if $v_C(0^-) = 0$.

Difficulty level   ○ Easy   ○ Normal   ● Hard
Calculation amount   ○ Small   ● Normal   ○ Large

1. $20e^{-t}u(t) - 10e^{-(t-2)}u(t-1) \, A$
2. $10e^{-t}u(t) - 10e^{-(t-1)}u(t-1) \, A$
3. $20e^{-t}u(t) - 20e^{-(t-1)}u(t-1) \, A$
4. $20e^{-t}u(t) + 20e^{-(t-1)}u(t-1) \, A$

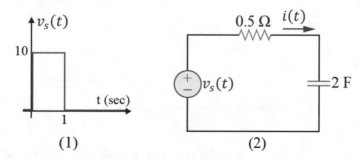

Figure 3.24 The circuit of problem 3.30

3.31. In the circuit of Figure 3.25, the primary current of each inductor is 2 A ($i_1(0^-) = i_2(0^-) = 2 \, A$), while the primary voltage of each capacitor is zero. Calculate the output voltage ($v_{out}(t)$) for $t \geq 0$.

Difficulty level   ○ Easy   ○ Normal   ● Hard
Calculation amount   ○ Small   ● Normal   ○ Large

1. $(2e^{-t} - 2te^{-t})$ V
2. $(2e^{-t} + 2te^{-t})$ V
3. $(-2e^{-t} + 2te^{-t} + 4e^{-2t})$ V
4. $(-2e^{-t} - 2te^{-t} + 4e^{-2t})$ V

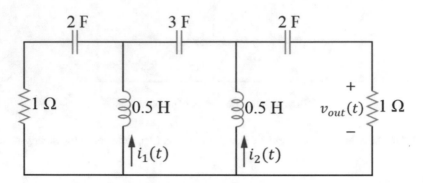

Figure 3.25 The circuit of problem 3.31

3.32. Which one of the choices below is true for the circuit, seen from terminal A–B, shown in Figure 3.26?

Difficulty level   ○ Easy   ○ Normal   ● Hard
Calculation amount   ○ Small   ○ Normal   ● Large

1. It is equivalent to a short circuit.
2. It is equivalent to a capacitor with the capacitance of α F.
3. It is equivalent to a resistor with the resistance of α Ω.
4. It is equivalent to an inductor with the inductance of α H.

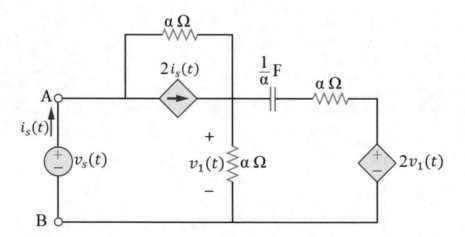

Figure 3.26 The circuit of problem 3.32

3.33. The circuit shown in Figure 3.27 has been in that situation for a long time. At $t = 0$, part of the circuit is cut down from the dashed line. Determine $v_{C1}(t)$ for $t > 0$.

Difficulty level   ○ Easy   ○ Normal   ● Hard
Calculation amount   ○ Small   ○ Normal   ● Large

1. $(4e^{-0.5t} + 6)$ V
2. $(4e^{-2t} + 6)$ V
3. $(8e^{-2t} + 2)$ V
4. $(8e^{-0.5t} + 2)$ V

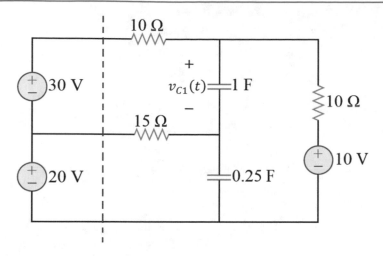

Figure 3.27 The circuit of problem 3.33

3.34. In the circuit of Figure 3.28, both switches are simultaneously closed. Calculate the voltage of 2-F capacitor exactly after the switching operation.

Difficulty level   ○ Easy   ○ Normal   ● Hard
Calculation amount ○ Small   ○ Normal   ● Large
1. 3 V
2. 4 V
3. 6 V
4. 9 V

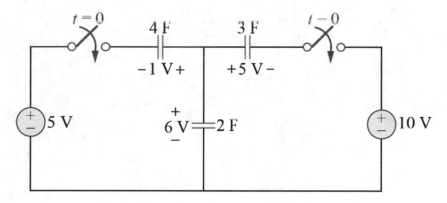

Figure 3.28 The circuit of problem 3.34

References

1. Rahmani-Andebili, M. (2020). DC Electrical circuit analysis: Practice problems, methods, and solutions, Springer Nature.
2. Rahmani-Andebili, M. (2020). AC Electrical circuit analysis: Practice problems, methods, and solutions, Springer Nature.

Solutions of Problems: Laplace Transform and Network Function

4

Abstract

In this chapter, the problems of the third chapter are fully solved, in detail, step-by-step, and with different methods.

4.1. The circuit of Figure 4.1.2 shows the main circuit in Laplace domain. The impedances of the components are as follows [1–2]:

$$Z_R = R \tag{1}$$

$$Z_C = \frac{1}{Cs} \tag{2}$$

Applying voltage division rule for the resistor:

$$V_{out}(s) = \frac{R}{R + \frac{1}{Cs}} \times V_{in}(s) \Rightarrow \frac{V_{out}(s)}{V_{in}(s)} = \frac{RCs}{RCs + 1}$$

Choice (2) is the answer.

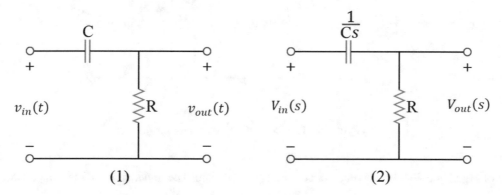

Figure 4.1 The circuit of solution of problem 4.1

4.2. The circuit of Figure 4.2.2 shows the main circuit in Laplace domain. The impedances of the components are as follows:

$$Z_R = R \Rightarrow Z_{1\,\Omega} = 1\,\Omega \tag{1}$$

$$Z_R = R \Rightarrow Z_{0.25\,\Omega} = 0.25\,\Omega \tag{2}$$

$$Z_C = \frac{1}{Cs} \Rightarrow Z_{2\,F} = \frac{1}{2s}\,\Omega \tag{3}$$

The right-side impedance ($\frac{1}{2s}\,\Omega$) does not have any effect on the input impedance seen from the left side of the circuit, since it is located on an open circuit branch. Therefore:

$$Z_{in} = 1 \left\| \left(\frac{1}{2s} + 0.25 \right) = \frac{1 \times \left(\frac{1}{2s} + 0.25 \right)}{1 + \left(\frac{1}{2s} + 0.25 \right)} = \frac{s+2}{5s+2}\,\Omega$$

Choice (1) is the answer.

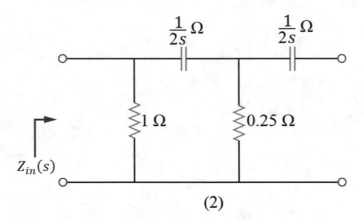

(1)

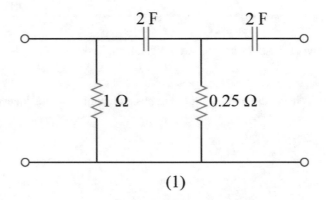

(2)

Figure 4.2   The circuit of solution of problem 4.2

4.3. The circuit of Figure 4.3.2 shows the main circuit in Laplace domain. The impedances of the components are as follows:

$$Z_R = R \Rightarrow Z_{2\,\Omega} = 2\,\Omega \tag{1}$$

$$Z_L = Ls \Rightarrow Z_{1\,H} = s\,\Omega \tag{2}$$

$$Z_C = \frac{1}{Cs} \Rightarrow Z_{1\,F} = \frac{1}{s}\,\Omega \tag{3}$$

Using voltage division rule for the output voltage:

$$V_{out}(s) = \frac{\frac{1}{s}}{\frac{1}{s}+s} V_{in}(s) = \frac{1}{s^2+1} V_{in}(s) \Rightarrow \frac{V_{out}(s)}{V_{in}(s)} = \frac{1}{s^2+1} \Rightarrow H(s) = \frac{V_{out}(s)}{V_{in}(s)} = \frac{1}{s^2+1}$$

Choice (1) is the answer.

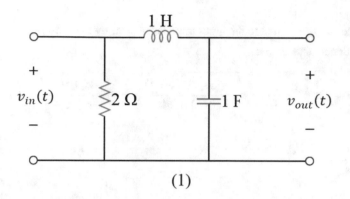

(1)

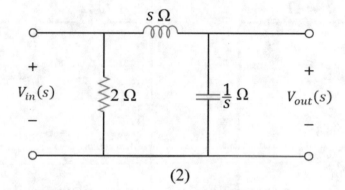

(2)

Figure 4.3 The circuit of solution of problem 4.3

4.4. Based on the information given in the problem, we have:

$$Y_{in}(s) = \frac{I(s)}{V(s)} = \frac{s^2+2s+3}{4s^3+5s^2+6s+7}$$

$$\Rightarrow (4s^3+5s^2+6s+7)I(s) = (s^2+2s+3)V(s)$$

Applying inverse Laplace transform:

$$\overset{L^{-1}}{\Longrightarrow} 4\frac{d^3}{dt^3}i(t) + 5\frac{d^2}{dt^2}i(t) + 6\frac{d}{dt}i(t) + 7i(t) = \frac{d^2}{dt^2}v(t) + 2\frac{d}{dt}v(t) + 3v(t)$$

Choice (3) is the answer.

4.5. The circuit of Figure 4.4.2 shows the main circuit in Laplace domain. The impedances of the components are as follows:

$$Z_L = Ls \Rightarrow Z_{1\,H} = s\,\Omega \tag{1}$$

$$Z_R = R \Rightarrow Z_{2\,\Omega} = 2\,\Omega \tag{2}$$

$$Z_R = R \Rightarrow Z_{4\,\Omega} = 4\,\Omega \tag{3}$$

$$Z_C = \frac{1}{Cs} \Rightarrow Z_{1\,F} = \frac{1}{s}\,\Omega \tag{4}$$

In this problem, the value of $\frac{V(s)}{I(s)}$ is equal to the input impedance seen by the voltage source. Therefore, we only need to calculate the input impedance as follows:

$$Z_{eq} = s + (s+2) \,\Big\|\, \left(4 + \frac{1}{s}\right) = s + \frac{(s+2)\left(4 + \frac{1}{s}\right)}{(s+2) + \left(4 + \frac{1}{s}\right)} = s + \frac{4s + 9 + \frac{2}{s}}{s + 6 + \frac{1}{s}} = s + \frac{4s^2 + 9s + 2}{s^2 + 6s + 1}$$

$$Z_{eq} = \frac{V(s)}{I(s)} = \frac{s^3 + 10s^2 + 10s + 2}{s^2 + 6s + 1}\,\Omega$$

Choice (2) is the answer.

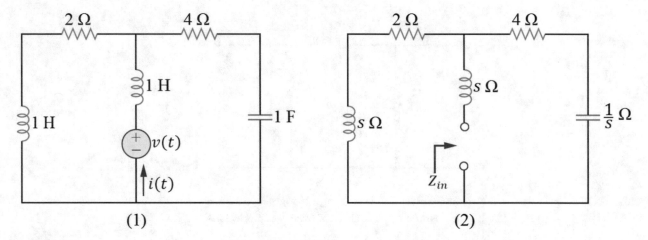

Figure 4.4   The circuit of solution of problem 4.5

4.6. The circuit of Figure 4.5.2 shows the main circuit in Laplace domain. The impedances of the components are as follows:

$$Z_C = \frac{1}{Cs} \Rightarrow Z_{0.5\,F} = \frac{1}{0.5s} = \frac{2}{s}\,\Omega \tag{1}$$

$$Z_R = R \Rightarrow Z_{3\,\Omega} = 3\,\Omega \tag{2}$$

$$Z_L = Ls \Rightarrow Z_{1\,H} = s\,\Omega \tag{3}$$

Applying current division rule:

$$I_L(s) = \frac{\frac{2}{s}}{\frac{2}{s} + 3 + s} I_s(s) = \frac{2}{s^2 + 3s + 2} I_s(s) \Rightarrow H(s) = \frac{I_L(s)}{I_s(s)} = \frac{2}{s^2 + 3s + 2}$$

Choice (2) is the answer.

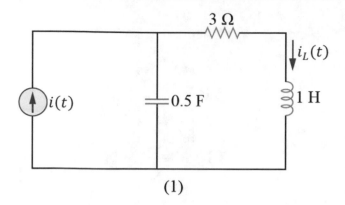

(1)

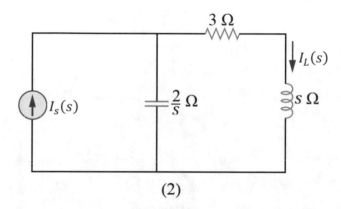

(2)

Figure 4.5 The circuit of solution of problem 4.6

4.7. The circuit of Figure 4.6.2 shows the main circuit in Laplace domain. The impedances of the components are as follows:

$$ Z_R = R \Rightarrow Z_{2\,\Omega} = 2\,\Omega \tag{1} $$

$$ Z_L = Ls \Rightarrow Z_{1\,H} = s\,\Omega \tag{2} $$

$$ Z_L = Ls \Rightarrow Z_{2\,H} = 2s\,\Omega \tag{3} $$

$$ Z_L = Ls \Rightarrow Z_{5\,H} = 5s\,\Omega \tag{4} $$

$$ Z_R = R \Rightarrow Z_{10\,\Omega} = 10\,\Omega \tag{5} $$

As is shown in Figure 4.6.2, to find the Thevenin impedance, we need to apply a test source (e.g., a test voltage source with the voltage and current of $V_t(s)$ and $I_t(s)$) to calculate the value of $\frac{V_t(s)}{I_t(s)}$.

Applying KVL in the left-side mesh:

$$ -V_t(s) + 2I_t(s) + sI_t(s) - 2sI(s) + sI_t(s) + 2sI(s) = 0 \Rightarrow -V_t(s) + 2(s+1)I_t(s) = 0 \Rightarrow \frac{V_t(s)}{I_t} = 2(s+1) $$

$$ \Rightarrow Z_{Th}(s) = 2(s+1)\,\Omega $$

Choice (3) is the answer.

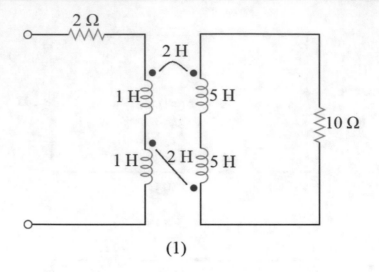

(1)

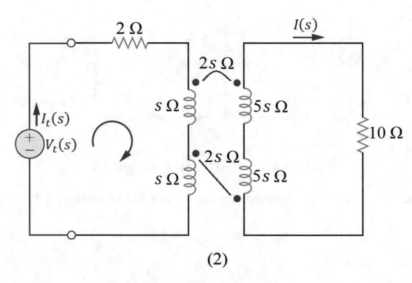

(2)

Figure 4.6 The circuit of solution of problem 4.7

4.8. Based on the information given in the problem, we know that:

$$v_{in}(t) = u(t) \overset{L}{\Rightarrow} V_{in}(s) = \frac{1}{s} \tag{1}$$

$$H(s) = \frac{V_{out}(s)}{V_{in}(s)} = \frac{25}{s^2 + 10s + 125} \Rightarrow V_{out}(s) = \frac{25}{s^2 + 10s + 125} V_{in}(s) \overset{(1)}{\Rightarrow} V_{out}(s) = \frac{25}{s(s^2 + 10s + 125)} \tag{2}$$

From initial value theorem, we know that:

$$f(0^+) = \lim_{s \to \infty} sF(s) \tag{3}$$

From final value theorem, we know that:

$$f(\infty) = \lim_{s \to 0} sF(s) \tag{4}$$

Solving (2) and (3):

$$v_{out}(0^+) = \lim_{s \to \infty} sV_{out}(s) = \lim_{s \to \infty} \frac{25}{(s^2 + 10s + 125)} = 0 \tag{5}$$

Solving (2) and (4):

$$v_{out}(\infty) = \lim_{s \to 0} sV_{out}(s) = \lim_{s \to 0} \frac{25}{(s^2 + 10s + 125)} = \frac{25}{125} = 0.2 \tag{6}$$

Choice (3) is the answer.

4.9. Based on the information given in the problem, we have:

$$h(t) = \left(e^{-t} - e^{-2t}\right)u(t) \xrightarrow{L} H(s) = \frac{1}{s+1} - \frac{1}{s+2} \tag{1}$$

$$x(t) = 2e^{-3t}u(t) \xrightarrow{L} X(s) = \frac{2}{s+3} \tag{2}$$

As we know, the network function (transfer function) is defined as follows:

$$H(s) = \frac{Y(s)}{X(s)} \tag{3}$$

Therefore:

$$Y(s) = H(s)X(s) \xrightarrow{(1),\,(?)} Y(s) = \left(\frac{1}{s+1} - \frac{1}{s+2}\right)\left(\frac{2}{s+3}\right) = \frac{2}{(s+1)(s+3)} - \frac{2}{(s+2)(s+3)}$$

$$\Rightarrow Y(s) = \frac{1}{s+1} - \frac{2}{s+2} + \frac{1}{s+3}$$

$$\xrightarrow{L^{-1}} y(t) = \left(e^{-t} - 2e^{-2t} + e^{-3t}\right)u(t)$$

Choice (2) is the answer.

4.10. Based on the information given in the problem, we have:

$$h(t) = \frac{3}{4}\left(e^{-t} + e^{-3t}\right)u(t) \xrightarrow{L} H(s) = \frac{3}{4}\left(\frac{1}{s+1} + \frac{1}{s+3}\right) \tag{1}$$

$$x(t) = 2\delta(t-5) \xrightarrow{L} X(s) = 2e^{-5s} \tag{2}$$

As we know, the network function (transfer function) is defined as follows:

$$H(s) = \frac{Y(s)}{X(s)} \tag{3}$$

Therefore:

$$Y(s) = H(s)X(s) \xrightarrow{(1),(2)} Y(s) = \frac{3}{4}\left(\frac{1}{s+1} + \frac{1}{s+3}\right) \times 2e^{-5s} = \frac{\frac{3}{2}}{s+1}e^{-5s} + \frac{\frac{3}{2}}{s+3}e^{-5s}$$

$$\xrightarrow{L^{-1}} y(t) = \frac{3}{2}\left(e^{-(t-5)} + e^{-3(t-5)}\right)u(t-5)$$

Choice (2) is the answer.

4.11. The circuit of Figure 4.7.2 shows the main circuit in Laplace domain. The impedances of the components are as follows:

$$Z_L = Ls \Rightarrow Z_{2\,H} = 2s\ \Omega \tag{1}$$

$$Z_C = \frac{1}{Cs} \Rightarrow Z_{0.5\,F} = \frac{1}{0.5s} = \frac{2}{s}\ \Omega \tag{2}$$

$$Z_R = R \Rightarrow Z_{1\,\Omega} = 1\ \Omega \tag{3}$$

$$Z_R = R \Rightarrow Z_{4\,\Omega} = 4\ \Omega \tag{4}$$

$$Z_L = Ls \Rightarrow Z_{0.5\,H} = 0.5s\ \Omega \tag{5}$$

$$Z_C = \frac{1}{Cs} \Rightarrow Z_{2\,F} = \frac{1}{2s}\ \Omega \tag{6}$$

Applying KCL in node 1:

$$\frac{V_1(s) - V_s(s)}{1} + \frac{V_1(s)}{0.5s + \frac{1}{2s}} + \frac{V_1(s) - V_2(s)}{4} = 0 \Rightarrow \left(\frac{5}{4} + \frac{2s}{s^2 + 1}\right)V_1(s) - \frac{1}{4}V_2(s) = V_s(s) \tag{7}$$

Applying KCL in node 2:

$$\frac{V_2(s) - V_1(s)}{4} + \frac{V_2(s)}{1} + \frac{V_2(s) - V_s(s)}{2s \left\| \frac{2}{s}\right.} = 0$$

$$\Rightarrow -\frac{1}{4}V_1(s) + \left(\frac{5}{4} + \frac{s^2 + 1}{2s}\right)V_2(s) = \left(\frac{s^2 + 1}{2s}\right)V_s(s) \tag{8}$$

Solving (7) and (8):

$$\frac{V_2(s)}{V_s(s)} = \frac{(s^2 + 1)}{(s + 1)^2}$$

Choice (4) is the answer.

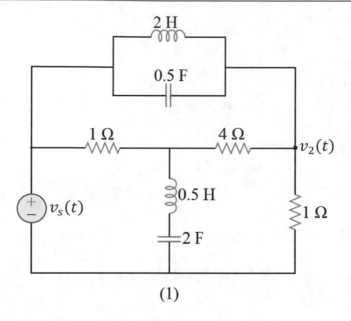

(1)

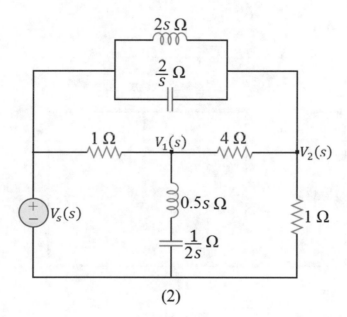

(2)

Figure 4.7   The circuit of solution of problem 4.11

4.12. Based on the information given in the problem, we know that the input $(i_s(t))$ is an impulse function. Therefore:

$$i_s(t) = \delta(t) \Rightarrow I_s(s) = 1 \tag{1}$$

The circuit of Figure 4.8.2 shows the main circuit in Laplace domain. The impedances of the components are as follows:

$$Z_R = R \Rightarrow Z_{4\,\Omega} = 4\,\Omega \tag{2}$$

$$Z_R = R \Rightarrow Z_{6\,\Omega} = 6\,\Omega \tag{3}$$

$$Z_L = Ls \Rightarrow Z_{1\,H} = s\,\Omega \tag{4}$$

$$Z_C = \frac{1}{Cs} \Rightarrow Z_{0.25\ F} = \frac{1}{0.25s} = \frac{4}{s}\ \Omega \tag{5}$$

Applying current division rule:

$$I(s) = \frac{4}{4 + Z_{box}} I_s(s) \tag{6}$$

where Z_{box} can be calculated as follows:

$$Z_{box} = (6+s) \left\| \frac{4}{s} = \frac{(6+s)\frac{4}{s}}{6+s+\frac{4}{s}} = \frac{4(s+6)}{s^2 + 6s + 4} \tag{7}$$

Solving (1), (6), and (7):

$$I(s) = \frac{4}{4 + \frac{4(s+6)}{s^2+6s+4}} \times 1 = \frac{s^2 + 6s + 4}{s^2 + 7s + 10} = 1 - \frac{s+6}{(s+2)(s+5)} = 1 + \frac{\frac{1}{3}}{s+5} - \frac{\frac{4}{3}}{s+2}$$

$$\xLeftarrow{L^{-1}} i(t) = \left(\frac{1}{3}e^{-5t} - \frac{4}{3}e^{-2t}\right)u(t) + \delta(t)$$

Choice (3) is the answer.

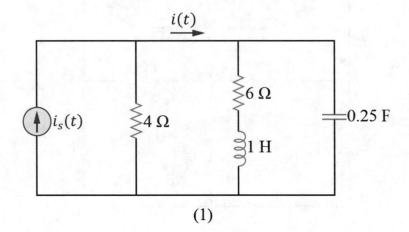

(1)

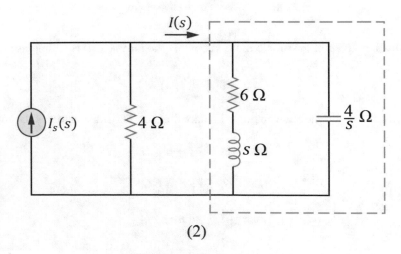

(2)

Figure 4.8 The circuit of solution of problem 4.12

4.13. The circuit of Figure 4.9.2 shows the main circuit in Laplace domain. The impedances of the components are as follows:

$$Z_R = R \Rightarrow Z_{1\,\Omega} = 1\ \Omega \tag{1}$$

$$Z_L = Ls \Rightarrow Z_{1\,H} = s\ \Omega \tag{2}$$

$$Z_C = \frac{1}{Cs} \Rightarrow Z_{1\,F} = \frac{1}{s}\ \Omega \tag{3}$$

The equivalent impedance seen by the current source can be calculated as follows:

$$Z_{eq} = 1 \left\| \left(s + \frac{1}{s} \,\right\| 1 \right)$$

$$= 1 \left\| \left(s + \frac{\frac{1}{s}}{\frac{1}{s} + 1} \right)\right.$$

$$= 1 \left\| \left(s + \frac{1}{s+1} \right)\right.$$

$$= 1 \left\| \left(\frac{s^2 + s + 1}{s + 1} \right)\right.$$

$$= \frac{\frac{s^2+s+1}{s+1}}{1 + \frac{s^2+s+1}{s+1}} = \frac{s^2 + s + 1}{s^2 + 2s + 2} \tag{4}$$

Using Ohm's law in the circuit of Figure 4.9.3:

$$V_a(s) = Z_{eq}I_s(s) \tag{5}$$

Solving (4) and (5):

$$V_a(s) = \frac{s^2 + s + 1}{s^2 + 2s + 2}I_s(s) \Rightarrow \left(s^2 + 2s + 2\right)V_a(s) = \left(s^2 + s + 1\right)I_s(s) \tag{6}$$

Applying inverse Laplace transform on (6):

$$\xrightarrow{L^{-1}} \frac{d^2}{dt^2}v_a(t) + 2\frac{d}{dt}v_a(t) + 2v_a(t) = \frac{d^2}{dt^2}i_s(t) + \frac{d}{dt}i_s(t) + i_s(t)$$

Choice (1) is the answer.

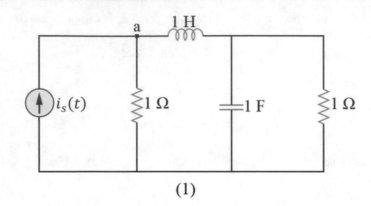

(1)

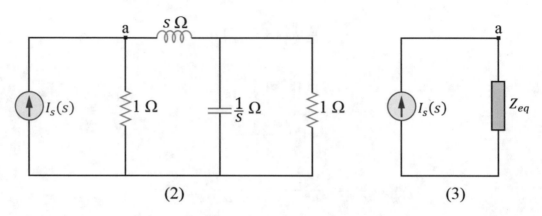

(2)   (3)

Figure 4.9 The circuit of solution of problem 4.13

4.14. The circuit of Figure 4.10.2 shows the main circuit in Laplace domain. The impedances of the components are as follows:

$$Z_R = R \Rightarrow Z_{1\,\Omega} = 1\,\Omega \tag{1}$$

$$Z_C = \frac{1}{Cs} \Rightarrow Z_{1\,F} = \frac{1}{s}\,\Omega \tag{2}$$

$$Z_L = Ls \Rightarrow Z_{1\,H} = s\,\Omega \tag{3}$$

The equivalent impedance, seen by the current source, is calculated as follows:

$$Z_{eq} = 1 + \left(1 \left\| \frac{1}{s}\right.\right) + s = 1 + \frac{\frac{1}{s}}{1 + \frac{1}{s}} + s = 1 + \frac{1}{s+1} + s = \frac{s^2 + 2s + 2}{s+1} \tag{4}$$

Using Ohm's law in the circuit of Figure 4.10.3:

$$V_s(s) = Z_{eq} I_s(s) \tag{5}$$

Solving (4) and (5):

$$V_s(s) = \frac{s^2 + 2s + 2}{s+1} I_s(s) \Rightarrow (s+1)V_s(s) = (s^2 + 2s + 2)I_s(s)$$

$$\xrightarrow{L^{-1}} \frac{d^2}{dt^2} i(t) + 2\frac{d}{dt} i(t) + 2i(t) = \frac{d}{dt} v_s(t) + v_s(t)$$

Choice (3) is the answer.

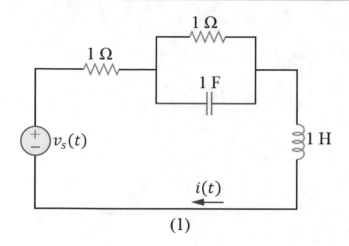

(1)

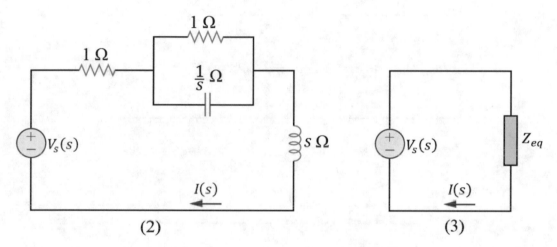

(2)   (3)

Figure 4.10 The circuit of solution of problem 4.14

4.15. The circuit of Figure 4.11.2 shows the primary circuit in Laplace domain. The impedances of the components are as follows:

$$Z_C = \frac{1}{Cs} \Rightarrow Z_{1\,F} = \frac{1}{s}\ \Omega \tag{1}$$

$$Z_R = R \Rightarrow Z_{1\,\Omega} = 1\ \Omega \tag{2}$$

Applying voltage division formula for the 1 Ω resistor in the vertical branch:

$$V(s) = \frac{1}{1 + \frac{1}{s}} \times V_{in}(s) = \frac{s}{s+1} \times V_{in}(s) \tag{3}$$

Applying KCL in the output node:

$$\frac{V_{out}(s) - 2V(s)}{1} + \frac{V_{out}(s) - V_{in}(s)}{\frac{1}{s}} = 0 \Rightarrow V_{out}(s) - 2V(s) + sV_{out}(s) - sV_{in}(s) = 0$$

$$\Rightarrow (s+1)V_{out}(s) - 2V(s) - sV_{in}(s) = 0 \tag{4}$$

Solving (3) and (4):

$$(s+1)V_{out}(s) - 2\frac{s}{s+1} \times V_{in}(s) - sV_{in}(s) = 0 \Rightarrow (s+1)V_{out}(s) - \left(\frac{2s}{s+1} + s\right)V_{in}(s) = 0$$

$$\Rightarrow (s+1)V_{out}(s) - \left(\frac{s^2+3}{s+1}\right)V_{in}(s) = 0 \Rightarrow \frac{V_{out}(s)}{V_{in}(s)} = \frac{s^2+3}{(s+1)^2}$$

Choice (1) is the answer.

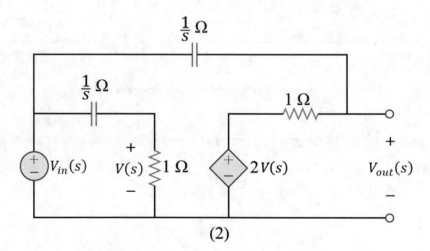

(1)

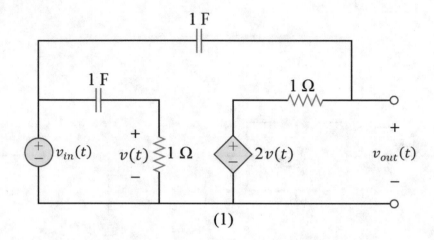

(2)

Figure 4.11 The circuit of solution of problem 4.15

4.16. The circuit of Figure 4.12.2 shows the primary circuit in Laplace domain. The impedances of the components are as follows:

$$Z_L = Ls \Rightarrow Z_{1\,H} = s\ \Omega \tag{1}$$

$$Z_L = Ls \Rightarrow Z_{2\,H} = 2s\ \Omega \tag{2}$$

$$Z_R = R \Rightarrow Z_{1\,\Omega} = 1\ \Omega \tag{3}$$

As is shown in Figure 4.12.3, to find the Thevenin impedance or the equivalent impedance in Laplace domain ($Z_{eq}(s)$), we need to turn off all the independent sources (herein, the independent voltage source must be replaced by a short circuit branch) and apply a test source (e.g., a test voltage source with the voltage and current of $V_t(s)$ and $I_t(s)$) to calculate the value of $\frac{V_t(s)}{I_t(s)}$ if the circuit includes a dependent source.

To simplify the problem, source transformation theorem is applied on the parallel connection of the dependent current source and the resistor to change it to the series connection of the dependent voltage source and the same resistor, as can be seen in Figure 4.12.3.

Applying KVL in the left-side mesh of the circuit of Figure 4.12.3:

$$sI(s) + s(I(s) + I_t(s)) + 2s(I(s) + I_t(s)) + sI(s) = 0 \Rightarrow 5sI(s) + 3sI_t(s) = 0$$

$$\Rightarrow I(s) = -\frac{3}{5}I_t(s) \tag{1}$$

$$-V_t(s) + I_t(s) \times 1 + 2I(s) + 2s(I(s) + I_t(s)) + sI(s) = 0$$

$$\Rightarrow -V_t(s) + (2s + 1)I_t(s) + (3s + 2)I(s) = 0 \tag{2}$$

Solving (1) and (2):

$$-V_t(s) + (2s + 1)I_t(s) + (3s + 2)\left(-\frac{3}{5}I_t(s)\right) = 0 \Rightarrow -V_t(s) + \frac{s-1}{5}I_t(s) = 0$$

$$\Rightarrow \frac{V_t(s)}{I_t(s)} = \frac{s-1}{5} \Rightarrow Z_{eq}(s) = \frac{s-1}{5}\ \Omega$$

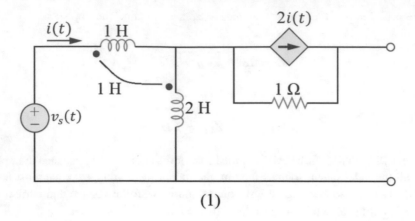

(1)

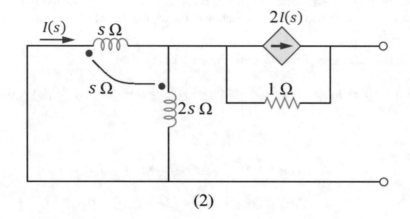

(2)

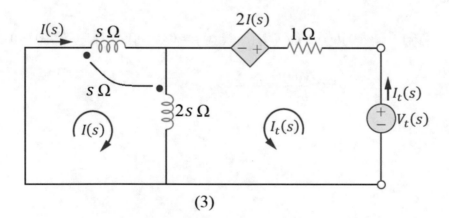

(3)

Figure 4.12 The circuit of solution of problem 4.16

4.17. The circuit of Figure 4.13.2 shows the main circuit in Laplace domain. The impedances of the components are as follows:

$$Z_R = R \Rightarrow Z_{2\,\Omega} = 2\,\Omega \tag{1}$$

$$Z_R = R \Rightarrow Z_{4\,\Omega} = 4\,\Omega \tag{2}$$

$$Z_C = \frac{1}{Cs} \Rightarrow Z_{2\,F} = \frac{1}{2s}\ \Omega \tag{3}$$

Applying voltage division rule for $V_2(s)$ in the circuit of Figure 4.13.2:

$$V_2(s) = \frac{\frac{1}{2s} + 4}{2 + \frac{1}{2s} + 4} \times V_{in}(s) = \frac{8s + 1}{12s + 1} V_{in}(s) \tag{4}$$

Applying voltage division rule for $V_1(s)$ in the circuit of Figure 4.13.2:

$$V_1(s) = \frac{2}{4 + \frac{1}{2s} + 2} \times V_{in}(s) = \frac{4s}{12s + 1} V_{in}(s) \tag{5}$$

$$V_{out}(s) = V_2(s) - V_1(s) = \frac{8s + 1}{12s + 1} V_{in}(s) - \frac{4s}{12s + 1} V_{in}(s) = \frac{4s + 1}{12s + 1} V_{in}(s)$$

$$\Rightarrow \frac{V_{out}(s)}{V_{in}(s)} = \frac{4s + 1}{12s + 1}$$

Choice (4) is the answer.

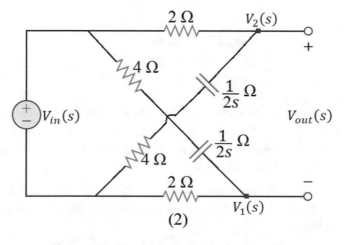

(1)

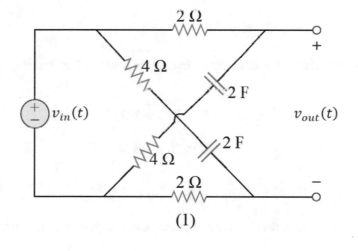

(2)

Figure 4.13 The circuit of solution of problem 4.17

4.18. Based on the information given in the problem, we have:

$$h(t) = \left(e^{-t} - e^{-3t}\right)u(t) \xrightarrow{L} H(s) = \frac{1}{s+1} - \frac{1}{s+3} \tag{1}$$

$$x(t) = 2e^{-2t} \xrightarrow{L} X(s) = \frac{2}{s+2} \tag{2}$$

As we know, the network function (transfer function) is defined as follows:

$$H(s) = \frac{Y(s)}{X(s)} \tag{3}$$

Therefore:

$$Y(s) = H(s)X(s) \xrightarrow{(1),(2)} Y(s) = \left(\frac{1}{s+1} - \frac{1}{s+3}\right)\frac{2}{s+2} = \frac{2}{(s+1)(s+2)} - \frac{2}{(s+2)(s+3)}$$

$$\Rightarrow Y(s) = \frac{2}{s+1} - \frac{4}{s+2} + \frac{2}{s+3}$$

$$\xrightarrow{L^{-1}} y(t) = \left(2e^{-t} - 4e^{-2t} + 2e^{-3t}\right)u(t)$$

Choice (3) is the answer.

4.19. The circuit of Figure 4.14.2 shows the main circuit in Laplace domain. The impedances of the components are as follows:

$$Z_R = R \Rightarrow Z_{2\,\Omega} = 2\,\Omega \tag{1}$$

$$Z_L = Ls \Rightarrow Z_{1\,H} = s\,\Omega \tag{2}$$

$$Z_C = \frac{1}{Cs} \Rightarrow Z_{0.5\,F} = \frac{1}{0.5s} = \frac{2}{s}\,\Omega \tag{3}$$

Moreover, the current of the independent current source in Laplace domain is as follows:

$$10u(t) \xrightarrow{L} \frac{10}{s} \tag{4}$$

Based on the information given in the problem, we have:

$$v_C(0^-) = 2\,V, i_L(0^-) = 1\,A \tag{5}$$

Since the capacitor has a nonzero primary voltage, it needs to be modeled by an impedance ($\frac{1}{Cs} = \frac{2}{s}\,\Omega$) in series with an independent voltage source ($\frac{v_C(0^-)}{s} = \frac{2}{s}$) in Laplace domain, as is shown in Figure 4.14.2. Likewise, the inductor has a nonzero primary current. Hence, it needs to be modeled by an impedance ($Ls = s\,\Omega$) in parallel with an independent current source ($\frac{i_L(0^-)}{s} = \frac{1}{s}$) in Laplace domain.

To simplify the circuit, source transformation theorem can be applied on the parallel connection of the independent current source and the inductor to change it to the series connection of the independent voltage source and the same inductor, as can be seen in Figure 4.14.3.

Now, by applying KVL in the right-side mesh, we have:

$$1 - sI_L(s) - 2I_L(s) + \left(\frac{10}{s} - I_L(s)\right)\frac{2}{s} + \frac{2}{s} = 0 \Rightarrow \left(-s - 2 - \frac{2}{s}\right)I_L(s) + 1 + \frac{10}{s} \times \frac{2}{s} + \frac{2}{s} = 0$$

$$\Rightarrow \left(s + 2 + \frac{2}{s}\right)I_L(s) = \left(1 + \frac{20}{s^2} + \frac{2}{s}\right) \Rightarrow I_L(s) = \frac{s^2 + 2s + 20}{s(s^2 + 2s + 2)} \ A$$

Choice (2) is the answer.

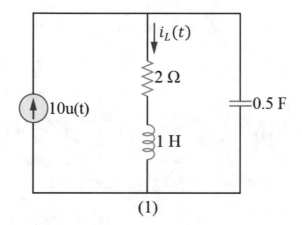

(1)

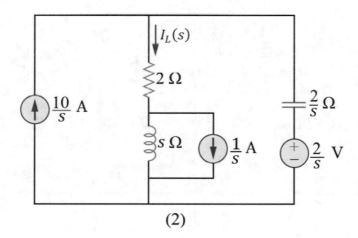

(2)

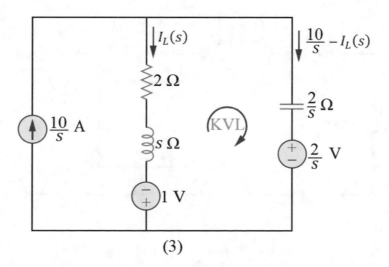

(3)

Figure 4.14 The circuit of solution of problem 4.19

4.20. The circuit of Figure 4.15.2 shows the main circuit in Laplace domain. The impedances of the components are as follows:

$$Z_L = Ls \Rightarrow Z_{1\,H} = s\ \Omega \tag{1}$$

$$Z_C = \frac{1}{Cs} \Rightarrow Z_{1\,F} = \frac{1}{s}\ \Omega \tag{2}$$

$$Z_R = R \Rightarrow Z_{1\,\Omega} = 1\ \Omega \tag{3}$$

Moreover, the voltage of the independent voltage source in Laplace domain is as follows:

$$v_s(s) = 5t \xrightarrow{L} V_s(s) = \frac{5}{s^2} \tag{4}$$

Applying voltage division rule:

$$V_{out}(s) = \frac{\left(1\,\middle\|\,\frac{1}{s}\right)}{s + \frac{1}{s} + 1 + \left(1\,\middle\|\,\frac{1}{s}\right)} V_s(s) \xrightarrow{(4)} V_{out}(s) = \frac{\frac{1 \times \frac{1}{s}}{1 + \frac{1}{s}}}{s + \frac{1}{s} + 1 + \frac{1 \times \frac{1}{s}}{1 + \frac{1}{s}}} \times \frac{5}{s^2} = \frac{\frac{1}{s+1}}{s + \frac{1}{s} + 1 + \frac{1}{s+1}} \times \frac{5}{s^2}$$

$$\Rightarrow V_{out}(s) = \frac{5s}{s^2(s^3 + 2s^2 + 3s + 1)}\ V$$

Choice (4) is the answer.

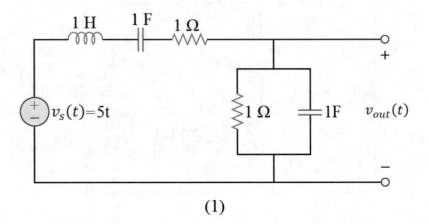

(1)

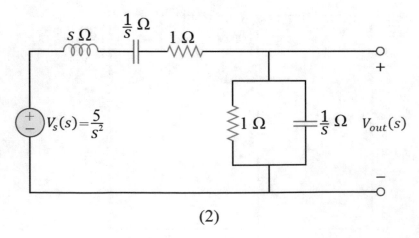

(2)

Figure 4.15 The circuit of solution of problem 4.20

4.21. The circuit of Figure 4.16.2 shows the main circuit in Laplace domain. The impedances of the components are as follows:

$$Z_R = R \Rightarrow Z_{2\,\Omega} = 2\,\Omega \tag{1}$$

$$Z_C = \frac{1}{Cs} \Rightarrow Z_{1\,F} = \frac{1}{s}\,\Omega \tag{2}$$

$$Z_C = \frac{1}{Cs} \Rightarrow Z_{2\,F} = \frac{1}{2s}\,\Omega \tag{3}$$

$$Z_R = R \Rightarrow Z_{1\,\Omega} = 1\,\Omega \tag{4}$$

Based on the information given in the problem, we know that the voltage source ($v_s(t)$) is an impulse function. Therefore:

$$v_s(t) = \delta(t) \overset{L}{\Longrightarrow} V_s(s) = 1 \tag{5}$$

From the circuit of Figure 4.16.2, it is clear that:

$$V_{out}(s) = 2V(s) \Rightarrow V(s) = \frac{1}{2}V_{out}(s) \tag{6}$$

Applying voltage division rule for the voltage of 1 Ω resistor:

$$V(s) = \frac{1}{1 + \frac{1}{2s}} V_A(s) \Rightarrow V(s) = \left(\frac{2s}{2s+1}\right) V_A(s) \Rightarrow V_A(s) = \left(\frac{2s+1}{2s}\right) V(s) \overset{(6)}{\Rightarrow} V_A(s) = \left(\frac{2s+1}{4s}\right) V_{out}(s) \tag{7}$$

Applying KCL in node A:

$$\frac{V_A(s) - V_s(s)}{2} + \frac{V_A(s)}{\frac{1}{s}} + \frac{V_A(s)}{\frac{1}{2s}+1} + \frac{V_A(s) - V_{out}(s)}{2} = 0$$

$$\Rightarrow \left(1 + s + \frac{2s}{2s+1}\right) V_A(s) - \frac{1}{2} V_s(s) - \frac{1}{2} V_{out}(s) = 0$$

$$\Rightarrow \left(\frac{2s^2 + 5s + 1}{2s+1}\right) V_A(s) - \frac{1}{2} V_s(s) - \frac{1}{2} V_{out}(s) = 0$$

$$\overset{(5),\,(7)}{\Longrightarrow} \left(\frac{2s^2 + 5s + 1}{2s+1}\right) \left(\frac{2s+1}{4s}\right) V_{out}(s) - \frac{1}{2} \times 1 - \frac{1}{2} V_{out}(s) = 0$$

$$\Rightarrow \left(\frac{2s^2 + 5s + 1}{4s} - \frac{1}{2}\right) V_{out}(s) - \frac{1}{2} = 0 \Rightarrow \left(\frac{2s^2 + 3s + 1}{4s}\right) V_{out}(s) = \frac{1}{2}$$

$$\Rightarrow V_{out}(s) = \frac{2s}{2s^2 + 3s + 1} = \frac{2s}{(s+1)(2s+1)} = \frac{2}{s+1} - \frac{2}{2s+1} = \frac{2}{s+1} - \frac{1}{s+\frac{1}{2}}$$

$$\overset{L^{-1}}{\Longrightarrow} v_{out}(t) = \left(2e^{-t} - e^{-0.5t}\right) u(t) \text{ V}$$

Choice (1) is the answer.

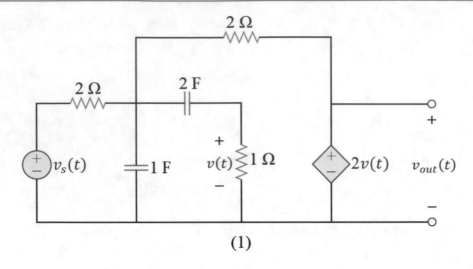

(1)

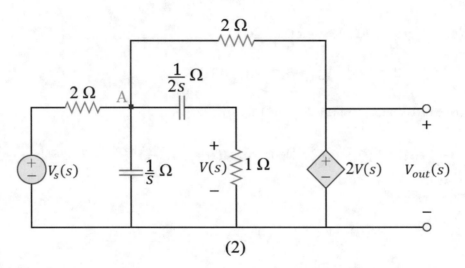

(2)

Figure 4.16 The circuit of solution of problem 4.21

4.22. The circuit has been presented in Laplace domain in Figure 4.17. Applying KCL in node A:

$$\frac{V_A(s) - V_1(s)}{s+1} + \frac{V_A(s)}{\frac{1}{s}} + \frac{V_A(s)}{s+1+\frac{1}{s}} = 0 \Rightarrow \left(\frac{1}{s+1} + s + \frac{s}{s^2+s+1}\right)V_A(s) - \left(\frac{1}{s+1}\right)V_1(s) = 0$$

$$\Rightarrow \frac{(s^2+s+1) + s(s^3+s^2+s+s^2+s+1) + (s^2+s)}{(s+1)(s^2+s+1)}V_A(s) = \frac{1}{s+1}V_1(s)$$

$$\Rightarrow \frac{s^4 + 2s^3 + 4s^2 + 3s + 1}{(s+1)(s^2+s+1)}V_A(s) = \frac{1}{s+1}V_1(s) \Rightarrow V_A(s) = \frac{s^2+s+1}{s^4+2s^3+4s^2+3s+1}V_1(s) \quad (1)$$

Applying voltage division rule for $V_2(s)$:

$$V_2(s) = \frac{\frac{1}{s}}{s+1+\frac{1}{s}}V_A(s) = \frac{1}{s^2+s+1}V_A(s) \Rightarrow V_A(s) = (s^2+s+1)V_2(s) \quad (2)$$

Solving (1) and (2):

$$\frac{s^2 + s + 1}{s^4 + 2s^3 + 4s^2 + 3s + 1} V_1(s) = (s^2 + s + 1) V_2(s) \Rightarrow \frac{V_2(s)}{V_1(s)} = \frac{1}{s^4 + 2s^3 + 4s^2 + 3s + 1}$$

Choice (2) is the answer.

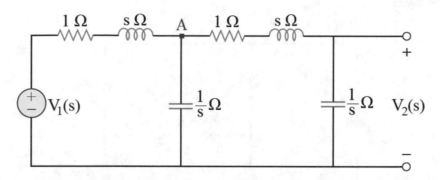

Figure 4.17 The circuit of solution of problem 4.22

4.23. The circuit of Figure 4.18.2 shows the main circuit in Laplace domain. The impedances of the components are as follows:

$$Z_{R_1} = R_1 \ \Omega \tag{1}$$

$$Z_L = Ls \ \Omega \tag{2}$$

$$Z_C = \frac{1}{Cs} \ \Omega \tag{3}$$

$$Z_{R_2} = R_2 \ \Omega \tag{4}$$

Based on the information given in the problem, we know that:

$$v_s(t) = A\cos(\omega t) \xrightarrow{L} V_s(s) = \frac{As}{s^2 + \omega^2} \tag{5}$$

Applying voltage division rule for node A:

$$V_A(s) = \frac{Ls}{Ls + R_1} V_s(s) = \frac{s}{s + \frac{R_1}{L}} V_s(s) \tag{6}$$

Applying voltage division rule for node B:

$$V_B(s) = \frac{R_2}{R_2 + \frac{1}{Cs}} V_s(s) = \frac{s}{s + \frac{1}{R_2 C}} V_s(s) \tag{7}$$

Therefore:

$$V_{out}(s) = V_A(s) - V_B(s) = \left(\frac{s}{s + \frac{R_1}{L}} - \frac{s}{s + \frac{1}{R_2 C}} \right) V_s(s) = \frac{\left(\frac{1}{R_2 C} - \frac{R_1}{L} \right) s}{s^2 + \left(\frac{R_1}{L} + \frac{1}{R_2 C} \right) s + \frac{R_1}{LR_2 C}} V_s(s) \tag{8}$$

Solving (5) and (8):

$$V_{out}(s) = \frac{\left(\frac{1}{R_2C} - \frac{R_1}{L}\right)s}{s^2 + \left(\frac{R_1}{L} + \frac{1}{R_2C}\right)s + \frac{R_1}{LR_2C}} \times \frac{As}{s^2 + \omega^2} = \frac{A\left(\frac{1}{R_2C} - \frac{R_1}{L}\right)s^2}{\left(s^2 + \left(\frac{R_1}{L} + \frac{1}{R_2C}\right)s + \frac{R_1}{LR_2C}\right)(s^2 + \omega^2)} V \tag{8}$$

Choice (2) is the answer.

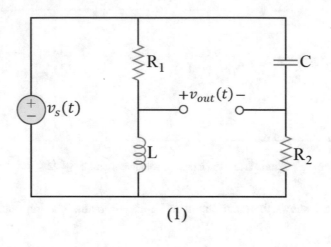

(1)

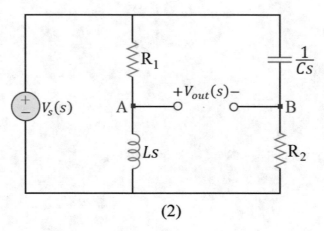

(2)

Figure 4.18 The circuit of solution of problem 4.23

4.24. Based on the information given in the problem, we have:

$$i(0^+) = 6\,A \tag{1}$$

$$Z_{in}(s) = \frac{s^2 + s + 2}{2s^2 + s + 1} \tag{2}$$

The circuit of Figure 4.19.2 shows the main circuit in Laplace domain. The voltage of the independent voltage source in Laplace domain is as follows:

$$v(t) = E \overset{L}{\Longrightarrow} V(s) = \frac{E}{s} \tag{3}$$

Using Ohm's law in Figure 4.19.2:

$$I(s) = \frac{V(s)}{Z_{in}(s)} = \frac{\frac{E}{s}}{\frac{s^2+s+2}{2s^2+s+1}} = \frac{E(2s^2+s+1)}{s(s^2+s+2)} \tag{4}$$

From initial value theorem, we know that:

$$f(0^+) = \lim_{s \to \infty} sF(s) \tag{5}$$

Solving (4) and (5):

$$i(0^+) = \lim_{s \to \infty} \frac{E(2s^2+s+1)}{(s^2+s+2)} = 2E \tag{6}$$

Solving (1) and (6):

$$2E = 6 \Rightarrow E = 3 \ V$$

Choice (1) is the answer.

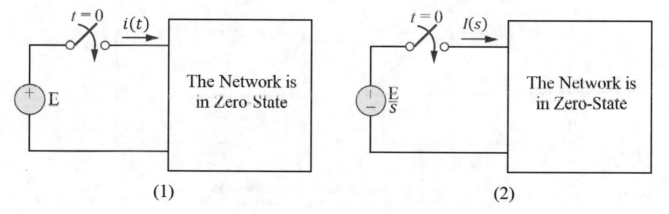

Figure 4.19 The circuit of solution of problem 4.24

4.25. The circuit of Figure 4.20.2 shows the main circuit in Laplace domain. The impedances of the components are as follows:

$$Z_L = Ls \tag{1}$$

$$Z_C = \frac{1}{Cs} \tag{2}$$

Moreover, $v_s(t)$ is a DC voltage source. Therefore, the voltage of the independent voltage source in Laplace domain is as follows:

$$v_s(t) = E \xrightarrow{L} V_s(s) = \frac{E}{s} \tag{3}$$

Applying voltage division rule:

$$V_C(s) = \frac{\frac{1}{Cs}}{\frac{1}{Cs} + Ls} V_s(s) \overset{(3)}{\Rightarrow} V_C(s) = \frac{\frac{E}{LC}}{s\left(s^2 + \frac{1}{LC}\right)} = E\left(\frac{1}{s} - \frac{s}{s^2 + \frac{1}{LC}}\right)$$

$$\overset{L^{-1}}{\Longrightarrow} v_C(t) = E\left(1 - \cos\left(\frac{1}{\sqrt{LC}}t\right)\right) \tag{4}$$

Based on the information given in the problem, the primary voltage of the capacitor and the primary current of the inductor are zero. In addition, the time (in second) that $v_C(t)$ is twice as $v_s(t)$ is requested. Thus:

$$v_C(t) = 2E \tag{5}$$

Solving (4) and (5):

$$E\left(1 - \cos\left(\frac{1}{\sqrt{LC}}t\right)\right) = 2E \Rightarrow 1 - \cos\left(\frac{1}{\sqrt{LC}}t\right) = 2 \Rightarrow \cos\left(\frac{1}{\sqrt{LC}}t\right) = -1 \Rightarrow \frac{1}{\sqrt{LC}}t = \pi$$

$$\Rightarrow t = \pi\sqrt{LC} \ sec$$

Choice (1) is the answer.

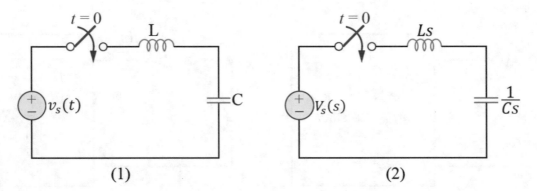

Figure 4.20 The circuit of solution of problem 4.25

4.26. The circuit of Figure 4.21.2 shows the primary circuit in Laplace domain. The impedances of the components are as follows:

$$Z_C = \frac{1}{Cs} \Rightarrow Z_{1 \, F} = \frac{1}{s} \, \Omega \tag{1}$$

$$Z_R = R \Rightarrow Z_{1 \, \Omega} = 1 \, \Omega \tag{2}$$

Moreover, the voltage of the independent voltage source and the current of the independent current source in Laplace domain are as follows:

$$V(s) = L\{\delta(t)\} = 1 \ V \tag{3}$$

$$I(s) = L\{\delta(t)\} = 1 \ A \tag{4}$$

To find the Thevenin equivalent circuit, we need to apply a test source (e.g., a test voltage source with the voltage and current of $V_t(s)$ and $I_t(s)$) to determine the relation between $V_t(s)$ and $I_t(s)$ in the form of $V_t(s) = \alpha I_t(s) + \beta$ [1]. Then, $Z_{Th}(s) = \alpha$ and $V_{Th}(s) = \beta$. Herein, the independent sources are not shut down.

Applying KCL in the supernode:

$$\frac{V_1(s) - 1}{\frac{1}{s}} + \frac{V_1(s)}{1} - 1 - I_t(s) = 0 \Rightarrow sV_1(s) - s + V_1(s) - 1 - I_t(s) = 0$$

$$\Rightarrow (s+1)V_1(s) - (s+1) - I_t(s) = 0 \tag{5}$$

Defining the voltage of the dependent voltage source based on the node voltages:

$$V_t(s) - V_1(s) = 2V_1(s) \Rightarrow V_1(s) = \frac{1}{3}V_t(s) \tag{6}$$

Solving (5) and (6):

$$(s+1)\frac{1}{3}V_t(s) - (s+1) - I_t(s) = 0 \Rightarrow V_t(s) = \frac{3}{s+1}I_t(s) + 3$$

$$\Rightarrow Z_{Th}(s) = \frac{3}{s+1} \ \Omega, \quad V_{Th}(s) = 3 \ V$$

Choice (1) is the answer.

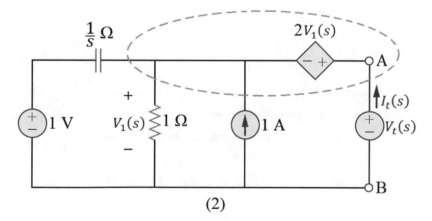

(1)

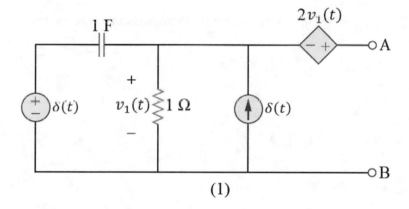

(2)

Figure 4.21 The circuit of solution of problem 4.26

4.27. Based on the information given in the problem, we have:

$$i_1(t) = \sqrt{2}u(t) \overset{L}{\Longrightarrow} I_1(s) = \frac{\sqrt{2}}{s} \tag{1}$$

$$H(s) = \frac{I_2(s)}{I_1(s)} = \frac{4(s+20)}{s+8} \Rightarrow I_2(s) = \frac{4(s+20)}{s+8} I_1(s) \tag{3}$$

Solving (1) and (3):

$$I_2(s) = \frac{4(s+20)}{s+8} \times \frac{\sqrt{2}}{s} \tag{4}$$

From final value theorem, we know that:

$$f(\infty) = \lim_{s \to 0} sF(s) \tag{5}$$

Solving (4) and (5):

$$i_2(\infty) = \lim_{s \to 0} sI_2(s) = \lim_{s \to 0} s \times \frac{4(s+20)}{s+8} \times \frac{\sqrt{2}}{s} = \lim_{s \to 0} \frac{4\sqrt{2}(s+20)}{s+8} = 10\sqrt{2} \tag{6}$$

Moreover, the value of energy, stored in an inductor, can be determined as follows:

$$W_L = \frac{1}{2} L(i_L)^2 \tag{7}$$

Solving (6) and (7):

$$W_L(\infty) = \frac{1}{2} L(i_2(\infty))^2 = \frac{1}{2} \times 1 \times 10^{-3} \times \left(10\sqrt{2}\right)^2 = 0.1\ J$$

Choice (1) is the answer.

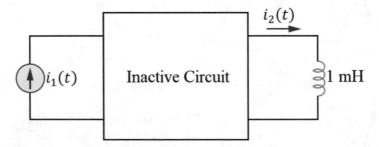

Figure 4.22 The circuit of solution of problem 4.27

4.28. The circuit of Figure 4.23.2 shows the main circuit in Laplace domain. The impedances of the components are as follows:

$$Z_R = R \Rightarrow Z_{1\ \Omega} = 1\ \Omega \tag{1}$$

$$Z_L = Ls \Rightarrow Z_{2\ H} = 2s\ \Omega \tag{2}$$

$$Z_C = \frac{1}{Cs} \tag{3}$$

Since the capacitor has a nonzero primary voltage, it is modeled by an impedance ($\frac{1}{Cs}$) in series with an independent voltage source ($\frac{v_C(0^-)}{s}$) in Laplace domain, as is shown in Figure 4.23.2.

Based on the information given in the problem, we have:

$$v_s(t) = \sin(2t)u(t) \Rightarrow V_s(s) = \frac{2}{s^2 + 4} \tag{4}$$

$$i_{out}(t > 0) = 0 \Rightarrow I_{out}(s) = 0 \tag{5}$$

Using Ohm's law for 1 Ω resistor:

$$V_1(s) = 1 \times I_{out}(s) = I_{out}(s) \overset{(2)}{\Rightarrow} V_1(s) = 0 \tag{6}$$

Applying KCL in the node 1:

$$-V_s(s) + \frac{V_1(s) - \frac{v_C(0^-)}{s}}{2s + \frac{1}{Cs}} + I_{out}(s) = 0 \tag{7}$$

Solving (4)–(7):

$$-\frac{2}{s^2+4} + \frac{0 - \frac{v_C(0^-)}{s}}{2s + \frac{1}{Cs}} + 0 = 0 \Rightarrow \frac{2}{s^2+4} = \frac{-\frac{v_C(0^-)}{s}}{2s + \frac{1}{Cs}} \Rightarrow \frac{2}{s^2+4} = \frac{-\frac{v_C(0^-)}{2}}{s^2 + \frac{1}{2C}} \Rightarrow \begin{cases} -\frac{v_C(0^-)}{2} = 2 \\ \frac{1}{2C} = 4 \end{cases}$$

$$\Rightarrow v_C(0^-) = -4\,V, C = \frac{1}{8}\,F$$

Choice (2) is the answer.

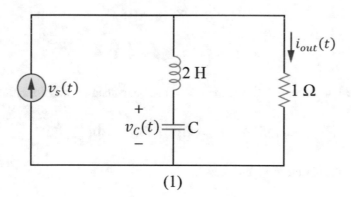

(1)

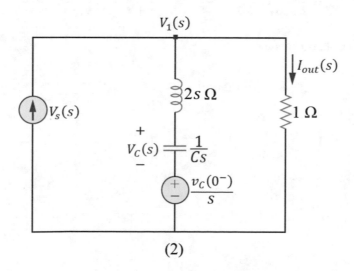

(2)

Figure 4.23 The circuit of solution of problem 4.28

4.29. Based on the information given in the problem, we have:

$$H(s) = \frac{V_{out}(s)}{V_{in}(s)} = \frac{10(s+1)}{s^2 + 2s + 3} \tag{1}$$

$$v_{in}(t) = 4\cos(2t) \xrightarrow{L} V_{in}(s) = \frac{4s}{s^2 + 4} \tag{2}$$

Solving (1) and (2):

$$\frac{V_{out}(s)}{\frac{4s}{s^2+4}} = \frac{10(s+1)}{s^2 + 2s + 3} \Rightarrow V_{out}(s) = \frac{10(s+1)}{s^2 + 2s + 3} \times \frac{4s}{s^2 + 4} \tag{4}$$

The steady-state response has been requested; therefore, we can use $s = j\omega = j2$, as $\omega = 2 \ rad/sec$ is the angular frequency of the power source. Hence:

$$V_{out}(\,j2) = \frac{10(\,j2 + 1)}{(\,j2)^2 + 2(\,j2) + 3} \times \frac{4(\,j2)}{(\,j2)^2 + 4} = 16.47 - j14.12 = 21.76e^{-j40.6}$$

$$\Rightarrow v_{out}(t) = 21.76\cos(2t - 40.6°)\ V$$

Choice (3) is the answer.

4.30. The circuit of Figure 4.24.3 shows the main circuit in Laplace domain. The impedances of the components are as follows:

$$Z_R = R \Rightarrow Z_{0.5\ \Omega} = 0.5\ \Omega \tag{1}$$

$$Z_C = \frac{1}{Cs} \Rightarrow Z_{2\ F} = \frac{1}{2s}\ \Omega \tag{2}$$

The pulse voltage, shown in the circuit of Figure 4.24.1, can be mathematically formulated as follows:

$$v_s(t) = 10(u(t) - u(t-1)) \xrightarrow{L} V_s(s) = 10\left(\frac{1}{s} - \frac{1}{s}e^{-s}\right) \tag{3}$$

Applying Ohm's law in the circuit of Figure 4.24.3:

$$I(s) = \frac{V_s(s)}{Z_{in}(s)} = \frac{10\left(\frac{1}{s} - \frac{1}{s}e^{-s}\right)}{0.5 + \frac{1}{2s}} = \frac{20(1 - e^{-s})}{s + 1} = \frac{20}{s + 1} - \frac{20e^{-s}}{s + 1} \tag{4}$$

Applying inverse Laplace transform on (4):

$$\xrightarrow{L^{-1}} i(t) = 20e^{-t}u(t) - 20e^{-(t-1)}u(t-1)\ A$$

Choice (3) is the answer.

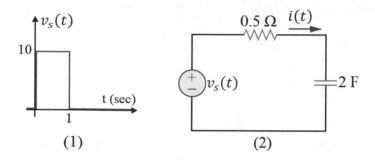

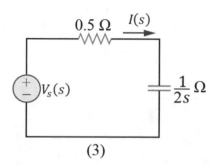

Figure 4.24 The circuit of solution of problem 4.30

4.31. The circuit of Figure 4.25.2 shows the main circuit in Laplace domain. The impedances of the components are as follows:

$$Z_R = R \Rightarrow Z_{1\,\Omega} = 1\,\Omega \tag{1}$$

$$Z_C - \frac{1}{Cs} \Rightarrow Z_{2\,F} = \frac{1}{2s}\,\Omega \tag{2}$$

$$Z_L = Ls \Rightarrow Z_{0.5\,H} = 0.5s\,\Omega \tag{3}$$

$$Z_C = \frac{1}{Cs} \Rightarrow Z_{3\,F} = \frac{1}{3s}\,\Omega \tag{4}$$

Based on the information given in the problem, we have:

$$i_1(0^-) = i_2(0^-) = 2\,A \tag{5}$$

Since each inductor has a nonzero primary current, each of them is modeled by an impedance ($Ls = 0.5s\,\Omega$) in parallel with an independent current source ($\frac{i_L(0^-)}{s} = \frac{2}{s}\,A$) in Laplace domain, as is shown in Figure 4.25.2.

To simplify the circuit, source transformation theorem can be applied on the parallel connection of the independent current source and the inductor to change it to the series connection of the independent voltage source and the same inductor, as can be seen in Figure 4.25.3.

By looking at the circuit of Figure 4.25.3, it is noticed that the circuit is symmetric. Therefore, the nodes A and B have equal voltages. Hence, no current flows through the middle capacitor. Thus, the circuit can be simplified, as is illustrated in Figure 4.25.4.

Applying voltage division rule for right-side 1 Ω resistor:

$$V_{out}(s) = \frac{1}{1 + 0.5s + \frac{1}{2s}} \times 1 \Rightarrow V_{out}(s) = \frac{2s}{s^2 + 2s + 1} = \frac{2}{s+1} - \frac{2}{(s+1)^2}$$

Applying inverse Laplace transform:

$$\overset{L^{-1}}{\Rightarrow} v_{out}(t) = (2e^{-t} - 2te^{-t})\ V$$

Choice (1) is the answer.

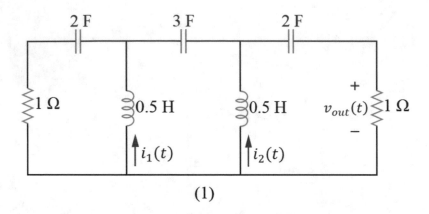

(1)

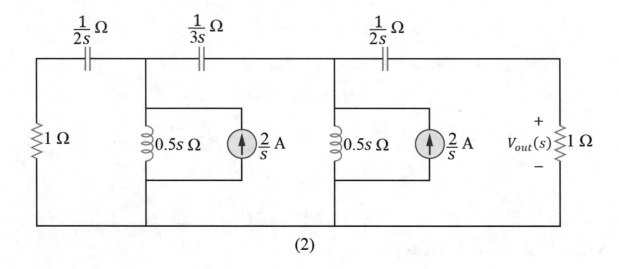

(2)

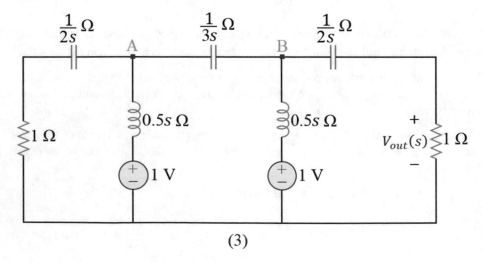

(3)

Figure 4.25 The circuit of solution of problem 4.31

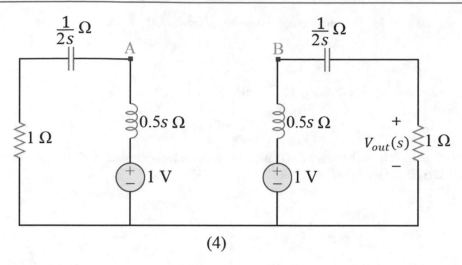

(4)

Figure 4.25 (continued)

4.32. The circuit of Figure 4.26.2 shows the main circuit in Laplace domain. The impedances of the components are as follows:

$$Z_R = R \Rightarrow Z_{\alpha\,\Omega} = \alpha\,\Omega \tag{1}$$

$$Z_C = \frac{1}{Cs} \Rightarrow Z_{\frac{1}{\alpha}\,F} = \frac{1}{\frac{1}{\alpha}s} = \frac{\alpha}{s}\,\Omega \tag{2}$$

Applying KCL in the supernode:

$$-I_s(s) + \frac{V_s(s) - V_1(s)}{\alpha} + 2I_s(s) = 0 \Rightarrow \frac{V_s(s) - V_1(s)}{\alpha} + I_s(s) = 0 \Rightarrow V_1(s) = V_s(s) + \alpha I_s(s) \tag{3}$$

Applying KCL in node A:

$$-2I_s(s) + \frac{V_1(s) - V_s(s)}{\alpha} + \frac{V_1(s)}{\alpha} + \frac{V_1(s) - 2V_1(s)}{\frac{\alpha}{s} + \alpha} = 0$$

$$\Rightarrow -2I_s(s) + \left(\frac{1}{\alpha} + \frac{1}{\alpha} - \frac{s}{\alpha(s+1)}\right)V_1(s) - \frac{1}{\alpha}V_s(s) = 0$$

$$\Rightarrow -2I_s(s) + \frac{s+2}{\alpha(s+1)}V_1(s) - \frac{1}{\alpha}V_s(s) = 0$$

$$\overset{(3)}{\Rightarrow} -2I_s(s) + \frac{s+2}{\alpha(s+1)}(V_s(s) + \alpha I_s(s)) - \frac{1}{\alpha}V_s(s) = 0$$

$$\Rightarrow \left(-2 + \frac{s+2}{s+1}\right)I_s(s) + \left(\frac{s+2}{\alpha(s+1)} - \frac{1}{\alpha}\right)V_s(s) = 0$$

$$\Rightarrow -\frac{s}{s+1}I_s(s) + \frac{1}{\alpha(s+1)}V_s(s) = 0 \Rightarrow \frac{s}{s+1}I_s(s) = \frac{1}{\alpha(s+1)}V_s(s)$$

$$\Rightarrow \frac{V_s(s)}{I_s(s)} = \frac{\frac{s}{s+1}}{\frac{1}{\alpha(s+1)}} = \alpha s \tag{4}$$

As we know, the value of $\frac{V_s(s)}{I_s(s)}$ presents the equivalent impedance seen by the source. Therefore:

$$\Rightarrow Z_{AB}(s) = \alpha s \tag{5}$$

Moreover, the impedance of an inductor can be determined as follows:

$$Z_L = Ls \tag{6}$$

By comparing (5) and (6), it can be concluded that the circuit, seen from terminal A–B, is equivalent to an inductor with the inductance of α H. Choice (4) is the answer.

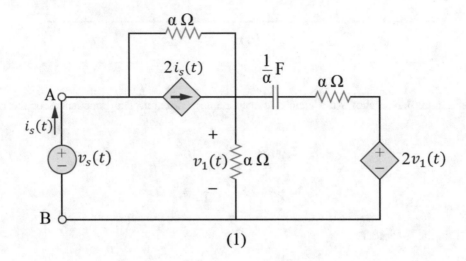

(1)

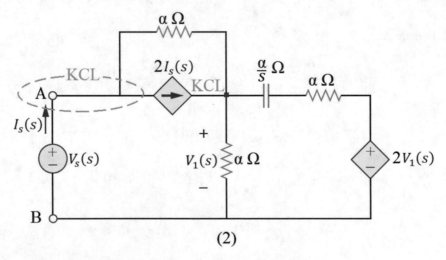

(2)

Figure 4.26   The circuit of solution of problem 4.32

4.33. Before $t = 0$, the circuit has reached its steady-state condition. Therefore, each capacitor is like an open circuit branch. Figure 4.27.2 shows the status of the primary circuit for $t = 0^-$.

Applying KVL in the indicated loop:

$$-20 - 30 + 10i(0^-) + 10i(0^-) + 10 = 0 \Rightarrow i(0^-) = 2\, A \tag{1}$$

Applying KVL in mesh 1:

$$-30 + 10i(0^-) + v_{C1}(0^-) + 15 \times 0 = 0 \Rightarrow v_{C1}(0^-) = 30 - 10i(0^-) \overset{(1)}{\Rightarrow} v_{C1}(0^-) = 10\, V \tag{2}$$

Applying KVL in mesh 2:

$$-20 + 15 \times 0 + v_{C2}(0^-) = 0 \Rightarrow v_{C2}(0^-) = 20 \tag{3}$$

Figure 4.27.3 shows the main circuit for $t = 0^+$ in Laplace domain. Each capacitor is modeled by an impedance in series with an independent voltage source ($\frac{v_C(0^-)}{s}$) in Laplace domain, since they have a nonzero primary voltage.

Moreover, the impedances of the components are as follows:

$$Z_R = R \Rightarrow Z_{10\,\Omega} = 10\, \Omega \tag{4}$$

$$Z_R = R \Rightarrow Z_{15\,\Omega} = 15\, \Omega \tag{5}$$

$$Z_C = \frac{1}{Cs} \Rightarrow Z_{1\,F} = \frac{1}{s}\, \Omega \tag{6}$$

$$Z_C = \frac{1}{Cs} \Rightarrow Z_{0.25\,F} = \frac{1}{0.25s} = \frac{4}{s}\, \Omega \tag{7}$$

Applying KVL in the indicated mesh in Figure 4.27.3:

$$10I(s) + \frac{10}{s} - \frac{20}{s} + \frac{4}{s}I(s) - \frac{10}{s} + \frac{1}{s}I(s) = 0 \Rightarrow I(s) = \frac{\frac{20}{s}}{10 + \frac{5}{s}} = \frac{4}{2s+1} \tag{8}$$

From the circuit, it is clear that:

$$V_{C1}(s) = -\frac{1}{s}I(s) + \frac{10}{s} \overset{(8)}{\Rightarrow} V_{C1}(s) = -\frac{1}{s}\left(\frac{4}{2s+1}\right) + \frac{10}{s} = -\frac{4}{s} + \frac{4}{s+\frac{1}{2}} + \frac{10}{s} = \frac{4}{s+\frac{1}{2}} + \frac{6}{s} \tag{9}$$

Applying inverse Laplace transform on (9):

$$\overset{L^{-1}}{\Rightarrow} v_{C1}(0^-) = \left(6 + 4e^{-0.5t}\right)\, V$$

Choice (1) is the answer.

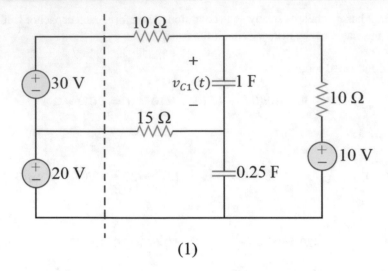

(1)

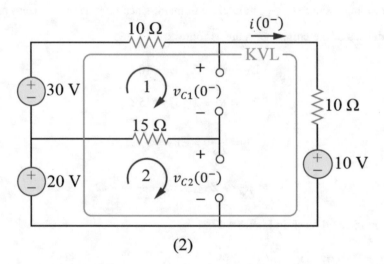

(2)

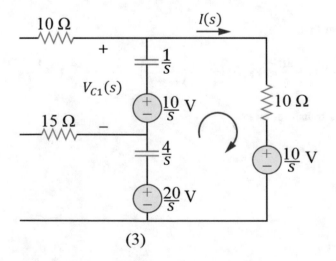

(3)

Figure 4.27 The circuit of solution of problem 4.33

4.34. The circuit of Figure 4.28.2 shows the main circuit in Laplace domain. The impedances of the components are as follows:

$$Z_C = \frac{1}{Cs} \Rightarrow Z_{4\,F} = \frac{1}{4s}\ \Omega \tag{1}$$

$$Z_C = \frac{1}{Cs} \Rightarrow Z_{3\,F} = \frac{1}{3s}\ \Omega \tag{2}$$

$$Z_C = \frac{1}{Cs} \Rightarrow Z_{2\,F} = \frac{1}{2s}\ \Omega \tag{3}$$

Moreover, the voltage of each independent voltage source in Laplace domain is as follows:

$$5\ V \xrightarrow{L} \frac{5}{s} \tag{4}$$

$$10\ V \xrightarrow{L} \frac{10}{s} \tag{5}$$

Since each capacitor has a nonzero primary voltage, they need to be modeled by an impedance ($\frac{1}{Cs}$) in series with an independent voltage source ($\frac{v_C(0^-)}{s}$) in Laplace domain, as is shown in Figure 4.28.2. Therefore:

$$1\ V \xrightarrow{L} \frac{1}{s}\ V \tag{6}$$

$$5\ V \xrightarrow{L} \frac{5}{s}\ V \tag{7}$$

$$6\ V \xrightarrow{L} \frac{6}{s}\ V \tag{8}$$

Applying KCL in the indicated node, which is for the voltage of 2 F capacitor:

$$\frac{V(s) - \frac{1}{s} - \frac{5}{s}}{\frac{1}{4s}} + \frac{V(s) - \frac{6}{s}}{\frac{1}{2s}} + \frac{V(s) - \frac{5}{s} - \frac{10}{s}}{\frac{1}{3s}} = 0$$

$$\Rightarrow 4s\left(V(s) - \frac{1}{s} - \frac{5}{s}\right) + 2s\left(V(s) - \frac{6}{s}\right) + 3s\left(V(s) - \frac{5}{s} - \frac{10}{s}\right) = 0$$

$$\Rightarrow 4sV(s) - 4 - 20 + 2sV(s) - 12 + 3sV(s) - 15 - 30 = 0 \Rightarrow 9sV(s) = 81 \Rightarrow V(s) = 9 \tag{9}$$

Applying inverse Laplace transform on (9):

$$\xrightarrow{L^{-1}} v(t) = 9u(t) \Rightarrow v(0^+) = 9\ V$$

Choice (4) is the answer.

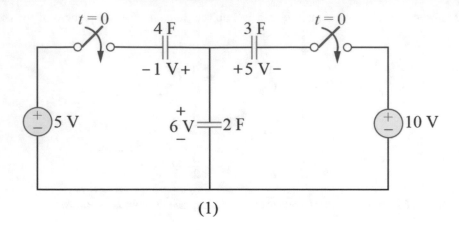

(1)

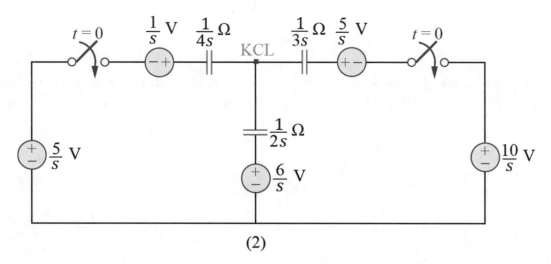

(2)

Figure 4.28 The circuit of solution of problem 4.34

References

1. Rahmani-Andebili, M. (2020). DC Electrical circuit analysis: Practice problems, methods, and solutions, Springer Nature.
2. Rahmani-Andebili, M. (2020). AC Electrical circuit analysis: Practice problems, methods, and solutions, Springer Nature.

Abstract

In this chapter, the basic and advanced problems concerned with the determination of natural frequencies of the electrical circuits are presented. The problems of this chapter include calculating the zero and nonzero natural frequencies of a circuit, determining the zeros and poles of the network function (transfer function) of a circuit, and identifying the damping status of a circuit. In this chapter, the problems are categorized in different levels based on their difficulty levels (easy, normal, and hard) and calculation amounts (small, normal, and large). Additionally, the problems are ordered from the easiest problem with the smallest computations to the most difficult problems with the largest calculations.

5.1. Determine the order and the number of nonzero natural frequencies of the circuit shown in Figure 5.1 [1–2].

Difficulty level   ● Easy   ○ Normal   ○ Hard
Calculation amount   ● Small   ○ Normal   ○ Large
1. 11 and 7
2. 11 and 10
3. 8 and 6
4. 8 and 8

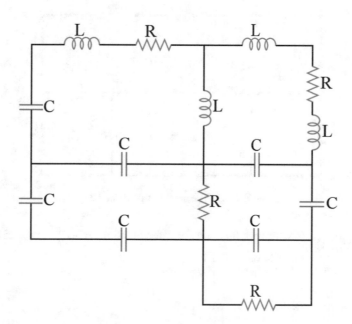

Figure 5.1 The circuit of problem 5.1

5.2. In the circuit of Figure 5.2, determine the number of state variables.

Difficulty level   ● Easy   ○ Normal   ○ Hard
Calculation amount   ● Small   ○ Normal   ○ Large
1. 3
2. 4
3. 5
4. 6

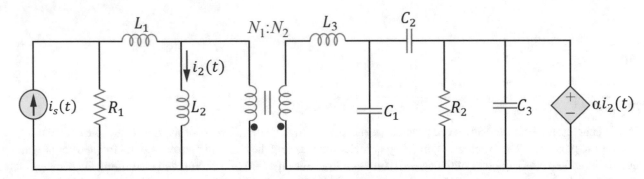

Figure 5.2 The circuit of problem 5.2

5.3. Which one of the following choices is true about the circuit of Figure 5.3?

Difficulty level   ● Easy   ○ Normal   ○ Hard
Calculation amount   ● Small   ○ Normal   ○ Large
1. It has 5 nonzero natural frequencies and 3 zero natural frequencies.
2. It has 4 nonzero natural frequencies and 3 zero natural frequencies.
3. It has 5 nonzero natural frequencies and 2 zero natural frequencies.
4. It has 4 nonzero natural frequencies and 2 zero natural frequencies.

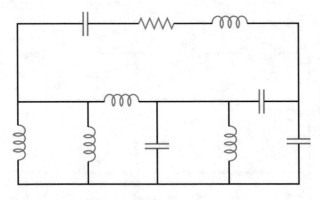

Figure 5.3 The circuit of problem 5.3

5.4. Determine the zero and poles of the network function (transfer function) of $\frac{V_{out}(s)}{V_s(s)}$ in the circuit of Figure 5.4.

Difficulty level   ○ Easy   ● Normal   ○ Hard
Calculation amount   ● Small   ○ Normal   ○ Large

1. $\begin{cases} \text{Poles} : 1 \pm j \\ \text{Zero} : -2 \end{cases}$

2. $\begin{cases} \text{Poles} : -1 \pm j \\ \text{Zero} : -2 \end{cases}$

3. $\begin{cases} \text{Poles} : 1 + j \\ \text{Zero} : -2 \end{cases}$

4. None of them

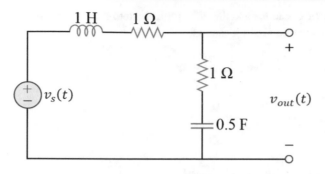

Figure 5.4 The circuit of problem 5.4

5.5. In the circuit of Figure 5.5, determine the zero and poles of the network function (transfer function), defined as $\frac{I_{out}(s)}{I_s(s)}$.

Difficulty level   ○ Easy   ● Normal   ○ Hard
Calculation amount   ● Small   ○ Normal   ○ Large

1. $\begin{cases} \text{Pole}: -30 \\ \text{Zero}: -10 \end{cases}$

2. $\begin{cases} \text{Pole}: -30 \\ \text{Zero}: 10 \end{cases}$

3. $\begin{cases} \text{Pole}: 30 \\ \text{Zero}: 10 \end{cases}$

4. None of them

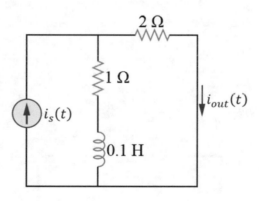

Figure 5.5 The circuit of problem 5.5

5.6. The state equations of a linear time-invariant (LTI) circuit is given in matrix form in the following. Determine the damping status of the circuit.

$$\begin{bmatrix} \dfrac{d}{dt} v_C(t) \\ \dfrac{d}{dt} i_L(t) \end{bmatrix} = \begin{bmatrix} 0 & 1 \\ -1 & -1 \end{bmatrix} \begin{bmatrix} v_C(t) \\ i_L(t) \end{bmatrix} + \begin{bmatrix} 1 \\ -1 \end{bmatrix} v_s(t)$$

Difficulty level   ○ Easy   ● Normal   ○ Hard
Calculation amount   ● Small   ○ Normal   ○ Large

1. Overdamped
2. Critically damped
3. Underdamped
4. Undamped

5.7. In the circuit of Figure 5.6, the capacitor and the inductor have nonzero primary voltage and current, respectively. Which one of the following choices is true about its status?

Difficulty level   ○ Easy   ● Normal   ○ Hard
Calculation amount   ○ Small   ● Normal   ○ Large

1. Overdamped
2. Critically damped
3. Underdamped
4. Undamped

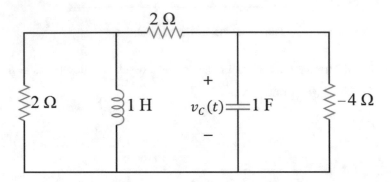

Figure 5.6 The circuit of problem 5.7

5.8. Which one of the following conditions can create an undamped status for the circuit shown in Figure 5.7?

Difficulty level   ○ Easy   ● Normal   ○ Hard
Calculation amount   ○ Small   ● Normal   ○ Large

1. $R = -\frac{1}{\sqrt{LC}}$
2. $R = -\frac{L}{rC}$
3. $R = -\frac{rC}{L}$
4. $R = -r$

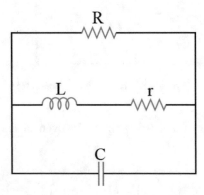

Figure 5.7 The circuit of problem 5.8

5.9. In a circuit, determine the poles of the network function (transfer function) of $\frac{V_{out}(s)}{V_{in}(s)}$ if the input and the output are as follows:

$$v_{in}(t) = 5e^{-2t}$$

$$v_{out}(t) = 3te^{-2t} + 2e^{-3t} \sin(6t)$$

Difficulty level   ○ Easy   ● Normal   ○ Hard
Calculation amount   ○ Small   ● Normal   ○ Large

1. $-2, \pm j6$
2. $2, 3, 0$
3. $2, 3, \pm j6$
4. $-2, -3 \pm j6$

5.10. For what value of α is the circuit of Figure 5.8 in the critically damped status?

Difficulty level   ○ Easy   ● Normal   ○ Hard
Calculation amount   ○ Small   ○ Normal   ● Large

1. -1.6
2. 0
3. 1
4. None of them

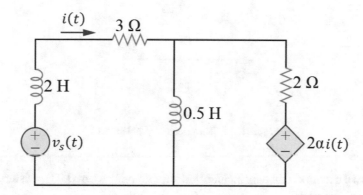

Figure 5.8 The circuit of problem 5.10

5.11. For what value of β is the circuit of Figure 5.9 in the undamped status?

Difficulty level   ○ Easy   ● Normal   ○ Hard
Calculation amount   ○ Small   ○ Normal   ● Large

1. 1
2. 2
3. 3
4. 4

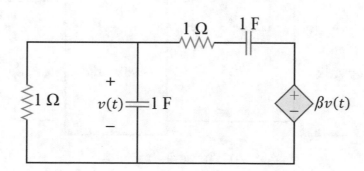

Figure 5.9 The circuit of problem 5.11

5.12. The zero-pole diagram concerned with the input impedance of a linear time-invariant (LTI) one-port network is shown in Figure 5.10. If the network is connected to a 1 A current source, 0.5 V can be measured across that after a while. Determine the input impedance of the network.

Difficulty level   ○ Easy   ○ Normal   ● Hard
Calculation amount   ● Small   ○ Normal   ○ Large

1. $\frac{2(s^2+1)}{(s+1)(s+2)}$ Ω

2. $\frac{(s^2+1)}{(s+1)(s+2)}$ Ω

3. $\frac{(s+1)(s+2)}{(s^2+1)}$ Ω

4. $\frac{(s+1)(s+2)}{2(s^2+1)}$ Ω

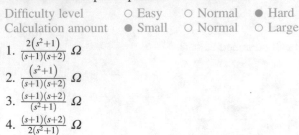

Figure 5.10 The circuit of problem 5.12

5.13. Figure 5.11 illustrates two networks that only include the linear time-invariant (LTI) resistors. If $i_s(t) = \cos(t) + \cos(2t)$, which one of the choices is correct?

Difficulty level   ○ Easy   ○ Normal   ● Hard
Calculation amount   ● Small   ○ Normal   ○ Large

1. $i_o(t)$ will be zero.
2. $v_o(t)$ will be zero.
3. $i_o(t)$ and $v_o(t)$ will be zero.
4. None of them.

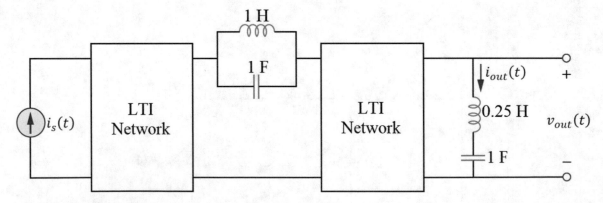

Figure 5.11 The circuit of problem 5.13

5.14. In the circuit of Figure 5.12, the input admittance has the zeros at $s = -2, -2.5$ and the poles at $s = -1, -1$. Determine the resistance of R_3.

Difficulty level   ○ Easy   ○ Normal   ● Hard
Calculation amount   ○ Small   ● Normal   ○ Large

1. $\frac{1}{4}$ Ω
2. $\frac{1}{8}$ Ω
3. 1 Ω
4. 2 Ω

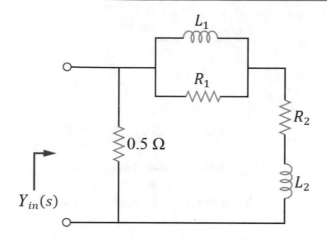

Figure 5.12 The circuit of problem 5.14

5.15. In the circuit of Figure 5.13, the input impedance has the poles at $s = -1, -3$ and the zeros at $s = -2, -4$. Determine the resistance of R_1.

Difficulty level   ○ Easy   ○ Normal   ● Hard
Calculation amount   ○ Small   ● Normal   ○ Large

1. $\frac{64}{55}$ Ω
2. $\frac{8}{5}$ Ω
3. 1 Ω
4. $\frac{8}{3}$ Ω

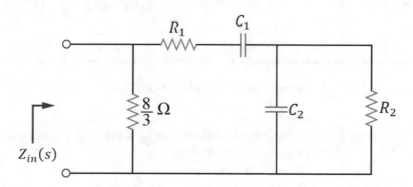

Figure 5.13 The circuit of problem 5.15

5.16. Figure 5.14 illustrates the result of a test carried out on the linear time-invariant (LTI) network, where:

$$i_{out}(t) = \left(2 - \frac{3}{2}e^{-\frac{1}{2}t}\right)u(t)$$
$$v_s(t) = u(t)$$

Which one of the following choices is true about the stability of the LTI network?

Difficulty level   ○ Easy   ○ Normal   ● Hard
Calculation amount   ○ Small   ● Normal   ○ Large

1. The LTI network is stable in both short circuit and open circuit statuses.
2. The LTI network is unstable in both short circuit and open circuit statuses.
3. The LTI network is stable in short circuit status but unstable in open circuit status.
4. The LTI network is stable in open circuit status but unstable in short circuit status.

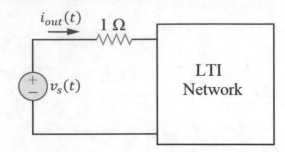

Figure 5.14 The circuit of problem 5.16

5.17. Which one of the inputs below will only show the natural frequencies in the output of the circuit of Figure 5.15?

Difficulty level   ○ Easy   ○ Normal   ● Hard
Calculation amount   ○ Small   ● Normal   ○ Large

1. $i_s(t) = e^{-1.5t}u(t)$ A
2. $i_s(t) = e^{-0.5t}u(t)$ A
3. $i_s(t) = e^{-t}u(t)$ A
4. $i_s(t) = e^{-2t}u(t)$ A

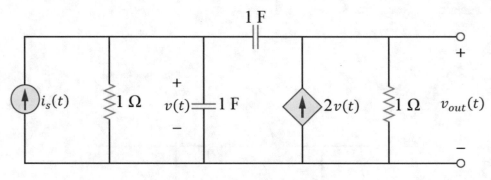

Figure 5.15 The circuit of problem 5.17

5.18. The location of zeros and poles of a network are shown in Figure 5.16. For what value of $a > 0$ will applying an input with the form of $e^{-at}u(t)$ not create an output in the same form?

Difficulty level   ○ Easy   ○ Normal   ● Hard
Calculation amount   ○ Small   ● Normal   ○ Large

1. 1
2. 2
3. 4
4. For no value of a

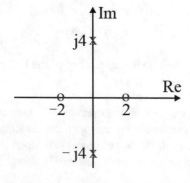

Figure 5.16 The circuit of problem 5.18

5.19. In the circuit of Figure 5.17, determine the nonzero natural frequencies.

1. $\frac{1}{3}\left(-1 \pm j\sqrt{5}\right)$
2. $\left(-1 \pm j\sqrt{5}\right),\ \pm j$
3. $\left(-1 \pm j\sqrt{5}\right),\ -1$
4. $\frac{1}{3}\left(-1 \pm j\sqrt{5}\right),\ \pm j,\ -1$

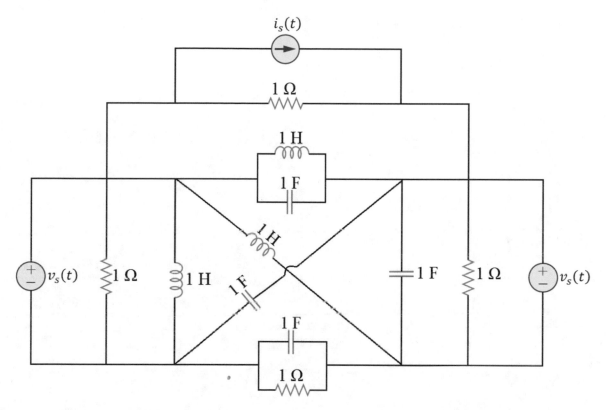

Figure 5.17 The circuit of problem 5.19

References

1. Rahmani-Andebili, M. (2020). DC Electrical circuit analysis: Practice problems, methods, and solutions, Springer Nature.
2. Rahmani-Andebili, M. (2020). AC Electrical circuit analysis: Practice problems, methods, and solutions, Springer Nature.

Solutions of Problems: Natural Frequencies of Electrical Circuits

6

Abstract

In this chapter, the problems of the fifth chapter are fully solved, in detail, step-by-step, and with different methods.

6.1. The number of energy-saving components (capacitors and inductors) of the circuit of Figure 6.1 is 11 [1–2]. However, there are two inductor cut-sets and one capacitor loop. Therefore, the number of the natural frequencies or the order of the circuit is $11 - 2 - 1 = 8$.

On the other hand, there are two capacitor cut-sets. Hence, two out of the eight natural frequencies are zero natural frequencies. Consequently, the number of nonzero natural frequencies are $8 - 2 = 6$.

Choice (3) is the answer.

© The Author(s), under exclusive license to Springer Nature Switzerland AG 2022
M. Rahmani-Andebili, *Advanced Electrical Circuit Analysis*, https://doi.org/10.1007/978-3-030-78540-6_6

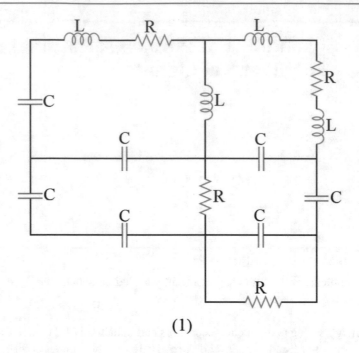

(1)

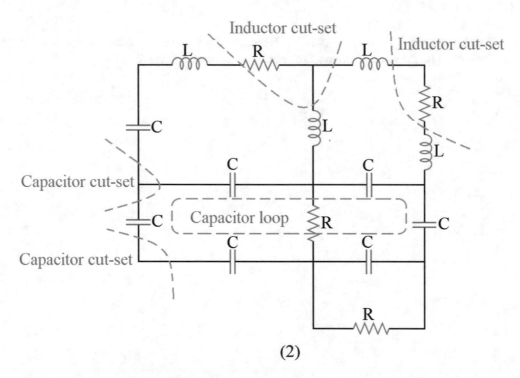

(2)

Figure 6.1 The circuit of solution of problem 6.1

6.2. The number of state variables of a circuit is equal to the number of energy-saving components (capacitors and inductors). However, as can be seen in Figure 6.2, there is one inductor cut-set and one capacitor loop in the circuit. Moreover, the voltage of C_3 depends on the current of L_2. Therefore, the number of state variables of the circuit is $6 - 1 - 1 - 1 = 3$.

Choice (1) is the answer.

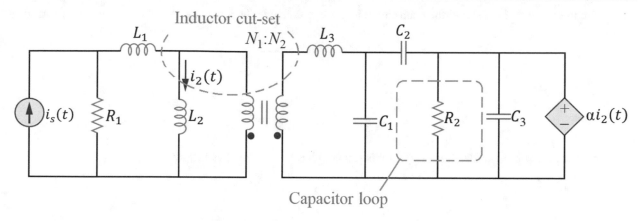

Figure 6.2 The circuit of solution of problem 6.2

6.3. The number of energy-saving components (capacitors and inductors) of the circuit of Figure 6.3 is nine. However, there is one inductor cut-set and one capacitor loop. Therefore, the number of the natural frequencies or the order of the circuit is $9 - 1 - 1 = 7$.

On the other hand, since there is one capacitor cut-set and two inductor loops, three out of the seven natural frequencies are zero natural frequencies. Consequently, the number of nonzero natural frequencies is four $(7 - 3 = 4)$. Choice (2) is the answer.

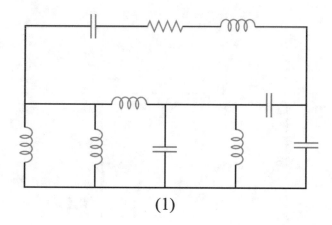

(1)

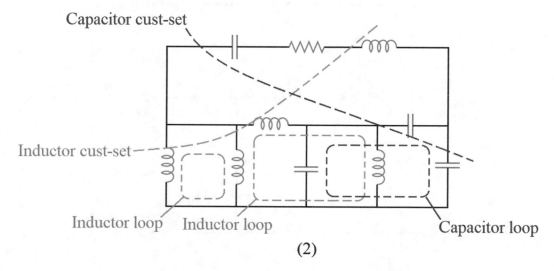

(2)

Figure 6.3 The circuit of solution of problem 6.3

6.4. The circuit of Figure 6.4.2 shows the main circuit in Laplace domain. The impedances of the components are as follows:

$$Z_L = Ls \Rightarrow Z_{1\,H} = s\,\Omega \tag{1}$$

$$Z_R = R \Rightarrow Z_{1\,\Omega} = 1\,\Omega \tag{2}$$

$$Z_C = \frac{1}{Cs} \Rightarrow Z_{0.5\,F} = \frac{1}{0.5s} = \frac{2}{s}\,\Omega \tag{3}$$

The network function (transfer function) for this problem has been defined as follows:

$$H(s) = \frac{V_{out}(s)}{V_s(s)} \tag{4}$$

Applying voltage division rule in the circuit of Figure 6.4.2:

$$V_{out}(s) = \frac{1 + \frac{2}{s}}{1 + \frac{2}{s} + s + 1}V_s(s) = \frac{s+2}{s^2 + 2s + 2}V_s(s) \Rightarrow H(s) = \frac{V_{out}(s)}{V_s(s)} = \frac{s+2}{s^2 + 2s + 2}$$

Therefore, the zero and poles of the network function (transfer function) are as follows:

$$\begin{cases} s^2 + 2s + 2 = 0 \Rightarrow s = \dfrac{-2 \pm \sqrt{2^2 - 4 \times 2}}{2} = -1 \pm j \Rightarrow Poles: \ -1 \pm j \\ s + 2 = 0 \Rightarrow s = -2 \Rightarrow Zero: \ -2 \end{cases}$$

Choice (2) is the answer.

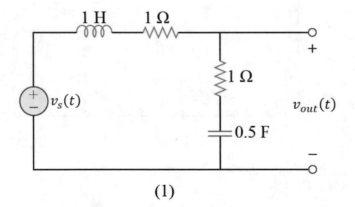

(1)

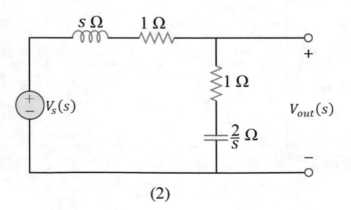

(2)

Figure 6.4 The circuit of solution of problem 6.4

6.5. The circuit of Figure 6.5.2 shows the main circuit in Laplace domain. The impedances of the components are as follows:

$$Z_R = R \Rightarrow Z_{1\,\Omega} = 1\,\Omega \tag{1}$$

$$Z_L = Ls \Rightarrow Z_{0.1\,H} = 0.1s\,\Omega \tag{2}$$

$$Z_R = R \Rightarrow Z_{2\,\Omega} = 2\,\Omega \tag{3}$$

The network function (transfer function) for this problem has been presented in the following form:

$$H(s) = \frac{I_{out}(s)}{I_s(s)} \tag{4}$$

Applying current division rule:

$$I_{out}(s) = \frac{1 + 0.1s}{1 + 0.1s + 2} I_s(s) = \frac{s + 10}{s + 30} I_s(s) \Rightarrow H(s) = \frac{I_{out}(s)}{I_s(s)} = \frac{s + 10}{s + 30}$$

Therefore, the zero and pole of the network function (transfer function) are as follows:

$$\begin{cases} s + 30 = 0 \Rightarrow s = -30 \Rightarrow Pole: \ -30 \\ s + 10 = 0 \Rightarrow s = -10 \Rightarrow Zero: \ -10 \end{cases}$$

Choice (1) is the answer.

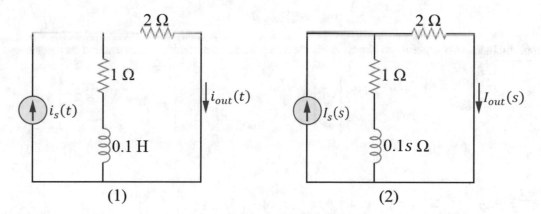

Figure 6.5 The circuit of solution of problem 6.5

6.6. The damping status of a circuit can be identified based on the roots of the characteristic equation of the circuit, where the characteristic equation is determined by solving the following equation:

$$\det(s[I] - [A]) = 0 \tag{1}$$

Based on the information given in the problem, we have:

$$[A] = \begin{bmatrix} 0 & 1 \\ -1 & -1 \end{bmatrix} \tag{2}$$

Solving (1) and (2):

$$\det\left(s\begin{bmatrix}1 & 0\\ 0 & 1\end{bmatrix} - \begin{bmatrix}0 & 1\\ -1 & -1\end{bmatrix}\right) = 0 \Rightarrow \det\left(\begin{bmatrix}s & -1\\ 1 & s+1\end{bmatrix}\right) = 0 \Rightarrow s(s+1) + 1 = 0$$

$$\Rightarrow s^2 + s + 1 = 0 \Rightarrow \begin{cases} b \neq 0\\ \Delta = 1^2 - 4 \times 1 = -3 < 0\end{cases}$$

Therefore, the damping status of the circuit is underdamped. Choice (3) is the answer.

6.7. The circuit of Figure 6.6.2 shows the main circuit in Laplace domain. The impedances of the components are as follows:

$$Z_R = R \Rightarrow Z_{2\,\Omega} = 2\,\Omega \tag{1}$$

$$Z_L = Ls \Rightarrow Z_{1\,H} = s\,\Omega \tag{2}$$

$$Z_C = \frac{1}{Cs} \Rightarrow Z_{1\,F} = \frac{1}{s}\,\Omega \tag{3}$$

$$Z_R = R \Rightarrow Z_{-4\,\Omega} = -4\,\Omega \tag{4}$$

As we know, the damping status of a circuit can be determined based on the roots of the characteristic equation of the circuit. Moreover, the characteristic equation of a circuit can be identified from the determinant of the nodal admittance matrix ($[Y_{nodal}]$) of the circuit, as follows:

$$\det([Y_{nodal}]) = 0 \tag{1}$$

The circuit includes two supernodes shown in Figure 6.6.2. The nodal admittance matrix can be determined as follows:

$$[Y_{nodal}] = \begin{bmatrix} \sum\limits_{j=1} y_{1j} & -y_{12}\\ -y_{21} & \sum\limits_{j=1} y_{2j}\end{bmatrix} = \begin{bmatrix} \dfrac{1}{2} + \dfrac{1}{s} + \dfrac{1}{2} & -\dfrac{1}{2}\\ -\dfrac{1}{2} & \dfrac{1}{-4} + s + \dfrac{1}{2}\end{bmatrix} = \begin{bmatrix} \dfrac{s+1}{s} & -\dfrac{1}{2}\\ -\dfrac{1}{2} & \dfrac{4s+1}{4}\end{bmatrix} \tag{2}$$

Solving (1) and (2):

$$\begin{vmatrix} \dfrac{s+1}{s} & -\dfrac{1}{2}\\ -\dfrac{1}{2} & \dfrac{4s+1}{4}\end{vmatrix} = 0 \Rightarrow \frac{s+1}{s} \times \frac{4s+1}{4} - \frac{1}{4} = 0 \Rightarrow \frac{4s^2 + 4s + 1}{4s} = 0 \Rightarrow (2s+1)^2 = 0 \Rightarrow s = -\frac{1}{2}, -\frac{1}{2}$$

Since the roots are real and equal, the circuit is in the critically damped status. Choice (2) is the answer.

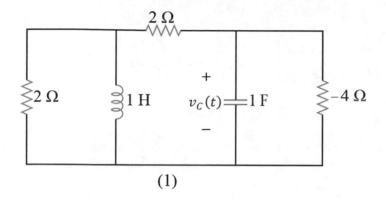

(1)

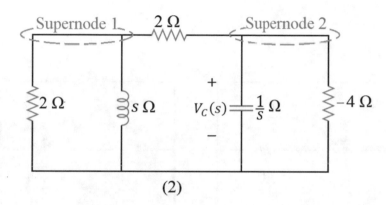

(2)

Figure 6.6 The circuit of solution of problem 6.7

6.8. The circuit of Figure 6.7.2 shows the main circuit in Laplace domain. The impedances of the components are as follows:

$$Z_R = R \tag{1}$$

$$Z_L = Ls \tag{2}$$

$$Z_r = r \tag{3}$$

$$Z_C = \frac{1}{Cs} \tag{4}$$

As we know, the damping status of a circuit can be determined based on the roots of the characteristic equation of the circuit. Moreover, the characteristic equation of a circuit can be identified from the determinant of the mesh impedance matrix ($[Z_{mesh}]$) of the circuit, as follows:

$$\det([Z_{mesh}]) = 0 \tag{1}$$

The circuit includes two meshes shown in Figure 6.7.2. The mesh impedance matrix can be determined as follows:

$$[Z_{mesh}] = \begin{bmatrix} \sum_{j=1} z_{1j} & -z_{12} \\ -z_{21} & \sum_{j=1} z_{2j} \end{bmatrix} = \begin{bmatrix} R + r + Ls & -(r + Ls) \\ -(r + Ls) & r + Ls + \frac{1}{Cs} \end{bmatrix} \tag{2}$$

Solving (1) and (2):

$$\begin{vmatrix} R+r+Ls & -(r+Ls) \\ -(r+Ls) & r+Ls+\dfrac{1}{Cs} \end{vmatrix} = 0 \Rightarrow (R+r+Ls)\left(r+Ls+\dfrac{1}{Cs}\right) - (r+Ls)(r+Ls) = 0$$

$$\Rightarrow RLCs^2 + (L+RrC)s + R + r = 0 \tag{3}$$

As we know, to create an undamped status, the roots must be on the imaginary axis. Therefore, the factor of s in (3) must be zero, as follows:

$$L + RrC = 0 \Rightarrow R = -\frac{L}{rC}$$

Choice (2) is the answer.

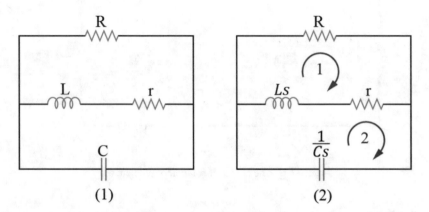

Figure 6.7 The circuit of solution of problem 6.8

6.9. Based on the information given in the problem, we have:

$$v_{in}(t) = 5e^{-2t} \overset{L}{\Rightarrow} V_{in}(s) = \frac{5}{s+2} \tag{1}$$

$$v_{out}(t) = 3te^{-2t} + 2e^{-3t}\sin(6t) \overset{L}{\Rightarrow} V_{out}(s) = -\frac{d}{ds}\left(\frac{3}{s+2}\right) + \frac{2 \times 6}{(s+3)^2 + 6^2} = \frac{3}{(s+2)^2} + \frac{12}{(s+3)^2 + 36} \tag{2}$$

Solving (1) and (2):

$$\frac{V_{out}(s)}{V_{in}(s)} = \frac{\frac{3}{(s+2)^2} + \frac{12}{(s+3)^2+36}}{\frac{5}{s+2}} = \frac{3(5s^2 + 22s + 61)}{5(s+2)\left((s+3)^2 + 36\right)}$$

$$\text{Poles} \Rightarrow 5(s+2)\left((s+3)^2 + 36\right) = 0 \Rightarrow s = -2, \ -3 \pm j6$$

Choice (4) is the answer.

6.10. The circuit of Figure 6.8.2 shows the main circuit in Laplace domain. The impedances of the components are as follows:

$$Z_L = Ls \Rightarrow Z_{2\,H} = 2s\ \Omega \tag{1}$$

$$Z_R = R \Rightarrow Z_{3\,\Omega} = 3\ \Omega \tag{2}$$

$$Z_L = Ls \Rightarrow Z_{0.5\,H} = 0.5s\ \Omega \tag{3}$$

$$Z_R = R \Rightarrow Z_{2\,\Omega} = 2\ \Omega \tag{4}$$

As we know, the damping status of a circuit can be determined based on the roots of the characteristic equation of the circuit. In addition, the characteristic equation of a circuit can be identified from the determinant of the mesh impedance matrix ($[Z_{mesh}]$) of the circuit, as follows:

$$\det([Z_{mesh}]) = 0 \tag{5}$$

The circuit includes two meshes illustrated in Figure 6.8.2. The mesh impedance matrix cannot be determined by using the straightforward rule, since there is a dependent source in the circuit.

KVL in the right-side mesh:

$$0.5s(I_2(s) - I(s)) + 2I_2(s) + 2\alpha I(s) = 0 \Rightarrow (-0.5s + 2\alpha)I(s) + (0.5s + 2)I_2(s) = 0 \tag{6}$$

KVL in the left-side mesh:

$$-V_s(s) + 2sI(s) + 3I(s) + 0.5s(I(s) - I_2(s)) = 0 \Rightarrow (2.5s + 3)I(s) - 0.5sI_2(s) = V_s(s) \tag{7}$$

Writing (6) and (7) in the form of $[Z][I] - [E_s]$.

$$\begin{bmatrix} -0.5s + 2\alpha & 0.5s + 2 \\ 2.5s + 3 & -0.5s \end{bmatrix} \begin{bmatrix} I(s) \\ I_2(s) \end{bmatrix} = \begin{bmatrix} 0 \\ V_s(s) \end{bmatrix} \Rightarrow [Z_{mesh}] = \begin{bmatrix} -0.5s + 2\alpha & 0.5s + 2 \\ 2.5s + 3 & -0.5s \end{bmatrix} \tag{8}$$

Solving (5) and (8):

$$\det\left(\begin{bmatrix} -0.5s + 2\alpha & 0.5s + 2 \\ 2.5s + 3 & -0.5s \end{bmatrix}\right) = 0 \Rightarrow (-0.5s + 2\alpha)(-0.5s) - (0.5s + 2)(2.5s + 3) = 0$$

$$\Rightarrow s^2 + (\alpha + 6.5)s + 6 = 0 \tag{9}$$

To put the circuit in the critically damped status, the discriminant of the characteristic equation must be zero. Hence:

$$(\alpha + 6.5)^2 - 4 \times 1 \times 6 = 0 \Rightarrow \alpha^2 + 13\alpha + \frac{169}{4} - 24 = 0 \Rightarrow \alpha^2 + 13\alpha + \frac{73}{4} = 0$$

$$\Rightarrow \alpha = \frac{-13 \pm \sqrt{13^2 - 73}}{2} = \frac{-13 \pm 4\sqrt{6}}{2} \Rightarrow \alpha \approx -1.6, -11.4$$

Choice (1) is the answer.

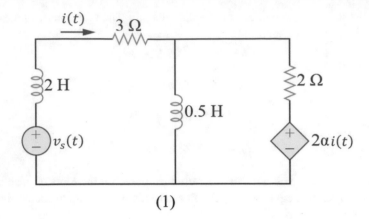

(1)

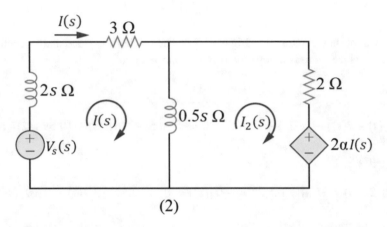

(2)

Figure 6.8 The circuit of solution of problem 6.10

6.11. The circuit of Figure 6.9.2 shows the main circuit in Laplace domain. The impedances of the components are as follows:

$$Z_R = R \Rightarrow Z_{1\,\Omega} = 1\,\Omega \tag{1}$$

$$Z_C = \frac{1}{Cs} \Rightarrow Z_{1\,F} = \frac{1}{s}\,\Omega \tag{2}$$

As we know, the damping status of a circuit can be determined based on the roots of the characteristic equation of the circuit. In addition, the characteristic equation of a circuit can be achieved from the determinant of the mesh impedance matrix ($[Z_{mesh}]$) of the circuit, as follows:

$$\det([Z_{mesh}]) = 0 \tag{3}$$

The circuit includes two meshes illustrated in Figure 6.9.2. The mesh impedance matrix cannot be determined by using the straightforward rule, since there is a dependent source in the circuit.

Defining $V(s)$ based on the mesh currents:

$$V(s) = \frac{1}{s}(I_1(s) - I_2(s)) \tag{4}$$

KVL in the right-side mesh:

$$\frac{1}{s}(I_2(s) - I_1(s)) + 1 \times I_2(s) + \frac{1}{s}I_2(s) + \beta V(s) = 0 \Rightarrow \left(-\frac{1}{s}\right)I_1(s) + \left(\frac{2}{s} + 1\right)I_2(s) + \beta V(s) = 0$$

$$\xrightarrow{\text{Using (4)}} \left(-\frac{1}{s}\right)I_1(s) + \left(\frac{2}{s} + 1\right)I_2(s) + \beta\frac{1}{s}(I_1(s) - I_2(s)) = 0 \Rightarrow \left(\frac{\beta - 1}{s}\right)I_1(s) + \left(\frac{2 - \beta}{s} + 1\right)I_2(s) = 0 \quad (5)$$

KVL in the left-side mesh:

$$1 \times I_1(s) + \frac{1}{s}(I_1(s) - I_2(s)) = 0 \Rightarrow \left(1 + \frac{1}{s}\right)I_1(s) - \frac{1}{s}I_2(s) = 0 \quad (6)$$

Writing (5) and (6) in the form of $[Z_{mesh}][I] = [E_s]$:

$$\begin{bmatrix} \frac{\beta - 1}{s} & \frac{2 - \beta}{s} + 1 \\ 1 + \frac{1}{s} & -\frac{1}{s} \end{bmatrix} \begin{bmatrix} I_1(s) \\ I_2(s) \end{bmatrix} = \begin{bmatrix} 0 \\ 0 \end{bmatrix} \Rightarrow [Z_{mesh}] = \begin{bmatrix} \frac{\beta - 1}{s} & \frac{2 - \beta}{s} + 1 \\ 1 + \frac{1}{s} & -\frac{1}{s} \end{bmatrix} \quad (7)$$

Solving (3) and (7):

$$\det\left(\begin{bmatrix} \frac{\beta - 1}{s} & \frac{2 - \beta}{s} + 1 \\ 1 + \frac{1}{s} & -\frac{1}{s} \end{bmatrix}\right) = 0 \Rightarrow \left(\frac{\beta - 1}{s}\right)\left(-\frac{1}{s}\right) - \left(\frac{2 - \beta}{s} + 1\right)\left(1 + \frac{1}{s}\right) = 0$$

$$\Rightarrow s^2 + (3 - \beta)s + 1 = 0 \quad (8)$$

To put the circuit in the undamped status, the roots of the characteristic equation must be on the imaginary axis. In other words, the value of b in the quadratic equation ($as^2 + bs + c = 0$) must be zero. Hence:

$$3 - \beta = 0 \rightarrow \beta = 3$$

Choice (3) is the answer.

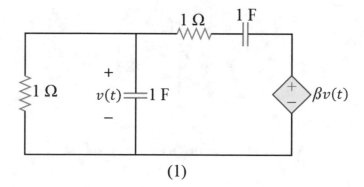

(1)

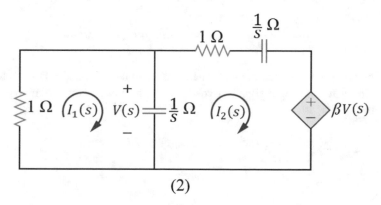

(2)

Figure 6.9 The circuit of solution of problem 6.11

6.12. Based on the zero-pole diagram of the input impedance of the one-port network, shown in Figure 6.10, we can determine the input impedance, as follows:

$$Z_{in}(s) = k\frac{(s-j)(s+j)}{(s+1)(s+2)} = \frac{k(s^2+1)}{(s+1)(s+2)} \tag{1}$$

Based on the information given in the problem, we know that if the network is connected to a 1 A current source, 0.5 V can be measured across that after a while. This means that the input impedance of the network for $s = 0$ (the input source is a DC source) is as follows:

$$Z_{in}(s) = \frac{V(s)}{I(s)} \Rightarrow Z_{in}(s = 0) = \frac{0.5}{1} = 0.5\ \Omega \tag{2}$$

Solving (1) for $s = 0$:

$$Z_{in}(s = 0) = \frac{k(0+1)}{(0+1)(0+2)} = \frac{k}{2} \tag{3}$$

Solving (2) and (3):

$$\frac{k}{2} = 0.5 \Rightarrow k = 1 \tag{4}$$

Solving (1) and (4):

$$Z_{in}(s) = \frac{(s^2+1)}{(s+1)(s+2)}\ \Omega$$

Choice (2) is the answer.

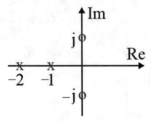

Figure 6.10 The circuit of solution of problem 6.12

6.13. The resonance frequency of the parallel LC circuit of the problem can be determined as follows:

$$\omega_{r1} = \frac{1}{\sqrt{LC}} \Rightarrow \omega_{r1} = \frac{1}{\sqrt{1\times 1}} = 1\ rad/sec$$

A parallel LC circuit behaves like an open circuit branch in its resonance frequency. Therefore, the parallel LC circuit will be like an open circuit for the input signal of cos(t). Thus, no voltage or current will be observed at the output terminal. In other words:

$$\omega_{r1} = 1\ rad/sec \Rightarrow v_{out}(t) = 0, i_{out}(t) = 0 \tag{1}$$

Moreover, the resonance frequency of the series LC circuit of the problem can be calculated as follows:

$$\omega_{r2} = \frac{1}{\sqrt{LC}} \Rightarrow \omega_{r2} = \frac{1}{\sqrt{\frac{1}{4} \times 1}} = 2 \; rad/\sec$$

A series LC circuit behaves like a short circuit branch in its resonance frequency. Therefore, the series LC circuit will be like a short circuit for the input signal of cos(2t). Hence:

$$\omega_{r2} = 2 \; rad/\sec \Rightarrow v_{out}(t) = 0 \tag{2}$$

From (1) and (2), it can be concluded that $v_{out}(t)$ is always zero for the input signal of $i_s(t) = \cos(t) + \cos(2t)$. Choice (2) is the answer.

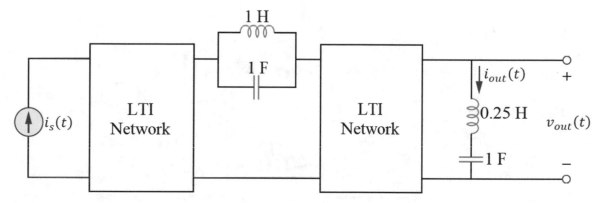

Figure 6.11 The circuit of solution of problem 6.13

6.14. Based on the information given in the problem, we know that the input admittance includes the zeros at $s = -2, -2.5$ and the poles at $s = -1, -1$. Therefore:

$$Y_{in}(s) = k\frac{(s+2)(s+2.5)}{(s+1)^2} \tag{1}$$

An inductor behaves like an open circuit at infinite frequency, since its impedance goes to infinity for $s = \infty$, as can be seen in the following:

$$Z_L(s) = Ls \Rightarrow Z_L(s = \infty) = L \times \infty \rightarrow \infty \tag{2}$$

The circuit of Figure 6.12.2 illustrates the main circuit for $s = \infty$. The input admittance of the circuit is as follows:

$$Y_{in}(s = \infty) = \frac{1}{0.5} = 2 \tag{3}$$

On the other hand, the value of input impedance for $s = \infty$ can be determined by using (1), as follows:

$$Y_{in}(s = \infty) = k \tag{4}$$

Solving (3) and (4):

$$k = 2 \tag{5}$$

An inductor behaves like a short circuit at zero frequency, since its impedance is zero for $s = 0$, as can be seen in the following:

$$Z_L(s) = Ls \Rightarrow Z_L(s = 0) = L \times 0 = 0 \tag{6}$$

The circuit of Figure 6.12.3 illustrates the main circuit for $s = 0$. The input admittance of the circuit is as follows:

$$Y_{in}(s = 0) = \left(0.5 \| R_2\right)^{-1} = \frac{0.5 + R_2}{0.5R_2} \tag{7}$$

On the other hand, the value of input impedance for $s = 0$ can be determined by using (1), as follows:

$$Y_{in}(s) = k\frac{(2)(2.5)}{(1)^2} = 5k \tag{8}$$

Solving (5), (7), and (8):

$$\frac{0.5 + R_2}{0.5R_2} = 5 \times 2 \Rightarrow 0.5 + R_2 = 5R_2 \Rightarrow R_2 = \frac{1}{8}\, \Omega$$

Choice (2) is the answer.

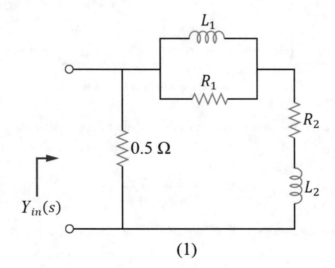

(1)

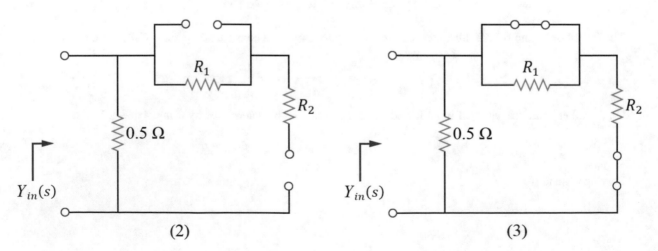

(2)   (3)

Figure 6.12   The circuit of solution of problem 6.14

6.15. Based on the information given in the problem, we know that the input impedance includes the poles at $s = -1, -3$ and the zeros at $s = -2, -4$. Therefore:

$$Z_{in}(s) = k\frac{(s+2)(s+4)}{(s+1)(s+3)} \tag{1}$$

A capacitor behaves like an open circuit at zero frequency, since its impedance goes to infinity for $s = 0$, as can be seen in the following:

$$Z_C(s) = \frac{1}{Cs} \Rightarrow Z_C(s=0) = \frac{1}{C \times 0} \to \infty \tag{2}$$

The circuit of Figure 6.13.2 illustrates the main circuit for $s = 0$. The input impedance of the circuit is as follows:

$$Z_{in}(s=0) = \frac{8}{3} \ \Omega \tag{3}$$

On the other hand, the value of input impedance for $s = 0$ can be determined by using (1), as follows:

$$Z_{in}(s=0) = k\frac{(2)(4)}{(1)(3)} = k\frac{8}{3} \tag{4}$$

Solving (3) and (4):

$$k\frac{8}{3} = \frac{8}{3} \Rightarrow k = 1 \tag{5}$$

A capacitor behaves like a short circuit at infinite frequency, since its impedance approaches zero for $s - \infty$, as can be seen in the following:

$$Z_C(s) = \frac{1}{Cs} \Rightarrow Z_C(s=\infty) = \frac{1}{C \times \infty} \to 0 \tag{6}$$

The circuit of Figure 6.13.3 illustrates the main circuit for $s = \infty$. The input impedance of the circuit is as follows:

$$Z_{in}(s=\infty) = \frac{8}{3} \| R_1 = \frac{\frac{8}{3}R_1}{\frac{8}{3}+R_1} \tag{7}$$

On the other hand, the value of input impedance for $s = \infty$ can be determined by using (1) and (5), as follows:

$$Z_{in}(s=\infty) = k \stackrel{(5)}{\Rightarrow} Z_{in}(s=\infty) = 1 \tag{8}$$

Solving (7) and ((8):

$$\frac{\frac{8}{3}R_1}{\frac{8}{3}+R_1} = 1 \Rightarrow \frac{8}{3} + R_1 = \frac{8}{3}R_1 \Rightarrow \frac{8}{3} = \frac{5}{3}R_1 \Rightarrow R_1 = \frac{8}{5} \ \Omega$$

Choice (2) is the answer.

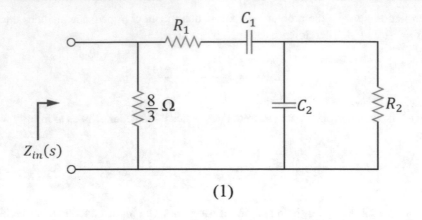

(1)

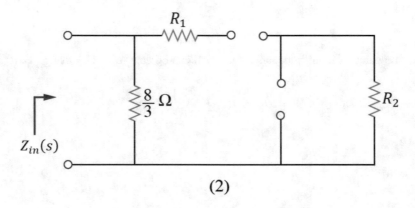

(2)

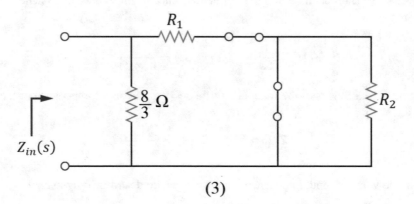

(3)

Figure 6.13 The circuit of solution of problem 6.15

6.16. The circuit of Figure 6.14.2 shows the main circuit in time domain. In Laplace domain, we have:

$$Z_R = R \Rightarrow Z_{1\,\Omega} = 1\,\Omega \tag{1}$$

$$v_s(t) = u(t) \overset{L}{\Rightarrow} V_s(s) = \frac{1}{s} \tag{2}$$

$$i_{out}(t) = \left(2 - \frac{3}{2}e^{-\frac{1}{2}t}\right)u(t) \overset{L}{\Rightarrow} I_{out}(s) = \frac{2}{s} - \frac{\frac{3}{2}}{s + \frac{1}{2}} \tag{3}$$

To determine the stability of the LTI network, we can determine the input admittance and the input impedance of the network, as follows:

$$Y_{in}(s) = \frac{I_{out}(s)}{V_s(s)} = \frac{\frac{2}{s} - \frac{\frac{3}{2}}{s + \frac{1}{2}}}{\frac{1}{s}} = \frac{s+2}{2s+1} \tag{4}$$

$$\Rightarrow Z_{in}(s) = \frac{1}{Y_{in}(s)} = \frac{2s+1}{s+2} \tag{5}$$

As can be noticed from Figure 6.14:

$$Z_N(s) = Z_{in}(s) - 1 = \frac{2s+1}{s+2} - 1 = \frac{s-1}{s+2} \tag{6}$$

$$\Rightarrow Y_N(s) = \frac{1}{Z_N(s)} = \frac{s+2}{s-1} \tag{7}$$

As can be noticed from (6), the input impedance of the network is stable, as its pole is in the left-half of the s-plane. Therefore, the LTI network is stable in open circuit status.

Moreover, as can be noticed from (7), the input admittance of the network is unstable, since its pole is in the right-half of the s-plane. Therefore, the LTI network is unstable in short circuit status.

Choice (4) is the answer.

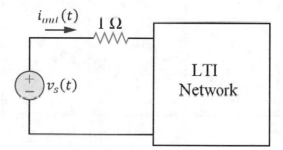

Figure 6.14 The circuit of solution of problem 6.16

6.17. The circuit of Figure 6.15.2 shows the main circuit in Laplace domain. The impedances of the components are as follows:

$$Z_R = R \Rightarrow Z_{1\,\Omega} = 1\,\Omega \tag{1}$$

$$Z_C = \frac{1}{Cs} \Rightarrow Z_{1\,F} = \frac{1}{s}\,\Omega \tag{2}$$

The network function (transfer function) for this problem is defined as follows:

$$H(s) = \frac{V_{out}(s)}{I_s(s)} \tag{3}$$

Applying KCL in the left-side supermesh in the circuit of Figure 6.15.2:

$$-I_s(s) + \frac{V(s)}{1} + \frac{V(s)}{\frac{1}{s}} + \frac{V(s) - V_{out}(s)}{\frac{1}{s}} = 0 \Rightarrow (2s+1)V(s) - sV_{out}(s) = I_s(s) \tag{4}$$

Applying KCL in the right-side supermesh circuit of Figure 6.15.2:

$$\frac{V_{out}(s) - V(s)}{\frac{1}{s}} - 2V(s) + \frac{V_{out}(s)}{1} = 0 \Rightarrow -(s+2)V(s) + (s+1)V_{out}(s) = 0$$

$$\Rightarrow V(s) = \frac{s+1}{s+2}V_{out}(s) \tag{5}$$

Solving (4) and (5):

$$(2s+1)\frac{s+1}{s+2}V_{out}(s) - sV_{out}(s) = I_s(s) \Rightarrow \frac{s^2+s+1}{s+2}V_{out}(s) = I_s(s) \quad (6)$$

Solving (3) and (6):

$$\Rightarrow H(s) = \frac{V_{out}(s)}{I_s(s)} = \frac{s+2}{s^2+s+1} \Rightarrow V_{out}(s) = \frac{s+2}{s^2+s+1}I_s(s) \quad (7)$$

As can be noticed from (7), if the input function ($I_s(s)$) has a pole at $s = -2$, this pole will be cancelled by the zero of the network function. Therefore, the natural frequency of the input signal will not be seen in the output ($V_{out}(s)$), and only the natural frequencies of the circuit will be observed. Hence:

$$I_s(s) = \frac{1}{s+2} \overset{L^{-1}}{\Rightarrow} i_s(t) = e^{-2t}u(t)\ A$$

Choice (4) is the answer.

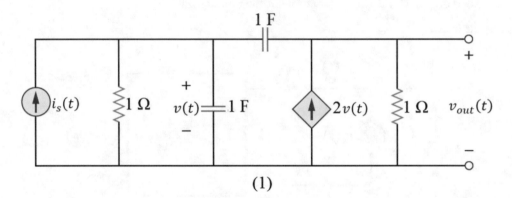

(1)

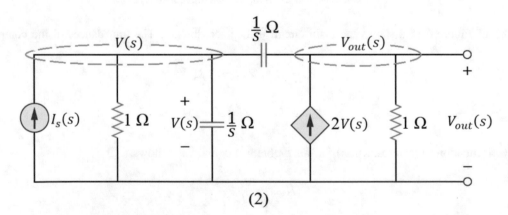

(2)

Figure 6.15 The circuit of solution of problem 6.17

6.18. Based on the zero-pole diagram of the network function (transfer function) of the circuit, shown in Figure 6.16, the network function can be determined, as follows:

$$H(s) = \frac{\text{Output function}}{\text{Input function}} = k\frac{(s+2)(s-2)}{(s+j4)(s-j4)} = \frac{k(s+2)(s-2)}{s^2+16}$$

$$\Rightarrow \text{Output function} = \frac{k(s+2)(s-2)}{s^2+16} \times \text{Input function} \quad (1)$$

As can be noticed from (1), if the input function has a pole at $s = -2$, this pole will be cancelled by the zero of the network function. Therefore, the natural frequency of the input function will not be seen in the output. Hence:

$$\text{Input function} = \frac{1}{s+2} \overset{L^{-1}}{\Rightarrow} \text{Input function in time domain} = e^{-2t}u(t) \tag{2}$$

Based on the information given in the problem, we know that:

$$\text{Input function in time domain} = e^{-at}u(t) \tag{3}$$

Solving (2) and (3):

$$a = 2$$

Choice (2) is the answer.

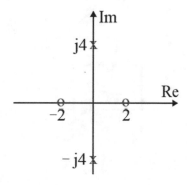

Figure 6.16 The circuit of solution of problem 6.18

6.19. To determine the natural frequencies of a circuit, we need to turn off the independent voltage and current sources, as can be seen in Figure 6.17.2.

By doing this, the parallel connection of the resistor and the inductor (on the left side of the circuit) is short-circuited and eliminated. Likewise, the parallel connection of the resistor and the capacitor (in the right side of the circuit) is removed, as is illustrated in Figure 6.17.3.

On the other hand, since the nonzero natural frequencies have been requested, we can simplify the circuit. Hence, the circuit of Figure 6.17.3 is simplified and shown in Figure 6.17.4, since all the components are in parallel.

Now, the resultant circuit is a parallel RLC circuit that its characteristic equation is as follows:

$$s^2 + 2\alpha s + \omega_0^2 = 0 \tag{1}$$

where

$$2\alpha = \frac{1}{RC} = \frac{1}{0.5 \times 3} = \frac{2}{3} \tag{2}$$

$$\omega_0^2 = \frac{1}{LC} = \frac{1}{0.5 \times 3} = \frac{2}{3} \tag{3}$$

Solving (1)–(3):

$$s^2 + \frac{2}{3}s + \frac{2}{3} = 0 \Rightarrow s = \frac{1}{3}\left(-1 \pm j\sqrt{5}\right)$$

Choice (1) is the answer.

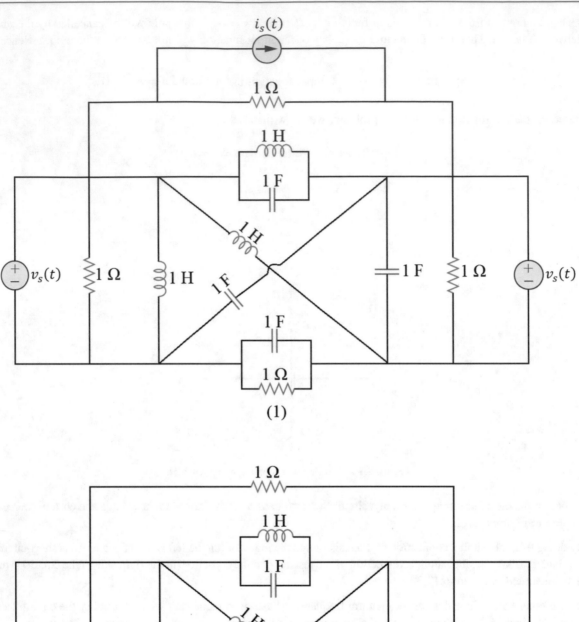

Figure 6.17 The circuit of solution of problem 6.19

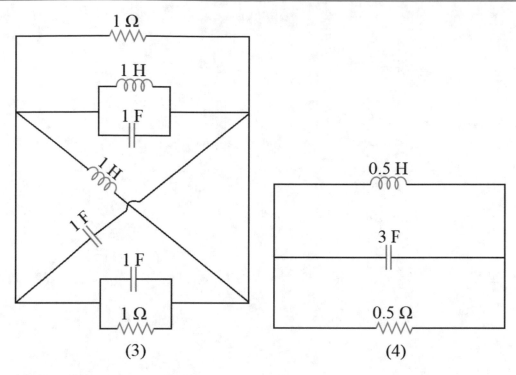

(3) (4)

Figure 6.17 (continued)

References

1. Rahmani-Andebili, M. (2020). DC Electrical circuit analysis: Practice problems, methods, and solutions, Springer Nature.
2. Rahmani-Andebili, M. (2020). AC Electrical circuit analysis: Practice problems, methods, and solutions, Springer Nature.

Problems: Network Theorems (Tellegen's and Linear Time-Invariant Network Theorems) 7

Abstract

In this chapter, the basic and advanced problems of network theorems, that is, Tellegen's and linear time-invariant (LTI) network theorems, are presented. In this chapter, the problems are categorized in different levels based on their difficulty levels (easy, normal, and hard) and calculation amounts (small, normal, and large). Additionally, the problems are ordered from the easiest problem with the smallest computations to the most difficult problems with the largest calculations.

7.1. The three-port network, shown in Figure 7.1, includes linear time-invariant (LTI) resistors and dependent sources [1–2]. The network is put under two tests as follows.

Test I: $v_s = 7\ V$, $i_s = 3\ A$, $i_{out} = 1\ A$
Test II: $v_s = 9\ V$, $i_s = 1\ A$, $i_{out} = 3\ A$
Determine the value of i_{out} for $v_s = 15\ V$, $i_s = 9\ A$

Difficulty level   ○ Easy   ● Normal   ○ Hard
Calculation amount   ● Small   ○ Normal   ○ Large
1. 0.3 A
2. 0.6 A
3. 3 A
4. 6 A

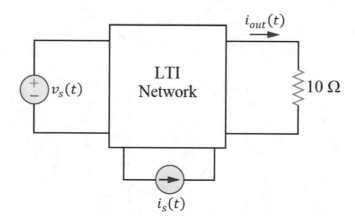

Figure 7.1 The circuit of problem 7.1

7.2. For the resistive linear time-invariant (LTI) two-port network, shown in Figure 7.2, the information below is given.

$$v_1(t) = 30t \ V, v_2(t) = 0, i_1(t) = 5t \ A, i_2(t) = 2t \ A$$

Determine the value of $i_1(t)$ in Ampere if $v_1(t) = 30t + 60 \ V$, $v_2(t) = 60t + 15 \ V$

Difficulty level   ○ Easy   ● Normal   ○ Hard
Calculation amount   ● Small   ○ Normal   ○ Large

1. $(t + 9) \ A$
2. $(9t + 11) \ A$
3. $(5t + 10) \ A$
4. $(-4t - 1) \ A$

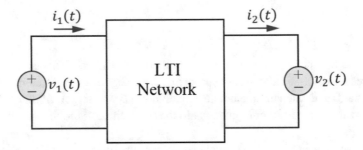

Figure 7.2 The circuit of problem 7.2

7.3. For the resistive linear time-invariant (LTI) two-port network, shown in Figure 7.3, the information below is given.

Test I: If $v_1 = 3 \ V$, $i_s = 3 \ A$, $i_{out} = 6 \ A$
Test II: $v_1 = 0 \ A$, $i_s = -2 \ A$, $i_{out} = 2 \ A$

Determine the value of $i_{out}(t)$ if $v_1(t) = -2 \ V$, $i_s(t) = 0 \ A$

Difficulty level   ○ Easy   ● Normal   ○ Hard
Calculation amount   ● Small   ○ Normal   ○ Large

1. $6 \ A$
2. $-4 \ A$
3. $4 \ A$
4. $-6 \ A$

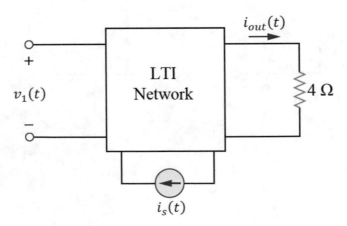

Figure 7.3 The circuit of problem 7.3

7.4. Figure 7.4.1 shows a resistive three-port network that a test is done on it. The same network is put under another test which is illustrated in Figure 7.4.2. Determine the value of $\tilde{v}_1(t)$.

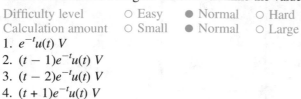

1. $e^{-t}u(t)$ V
2. $(t - 1)e^{-t}u(t)$ V
3. $(t - 2)e^{-t}u(t)$ V
4. $(t + 1)e^{-t}u(t)$ V

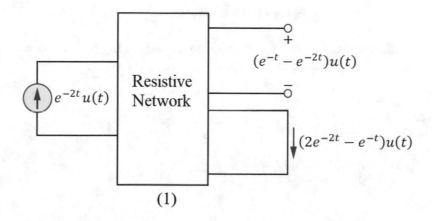

(1)

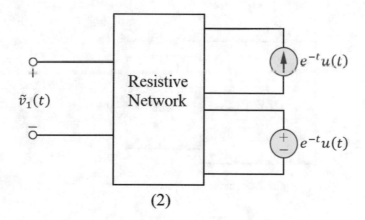

(2)

Figure 7.4 The circuit of problem 7.4

7.5. Figure 7.5 shows a linear time-invariant (LTI) two-port network that includes RLC components.

$$i_1(t) = 2\sin\left(\omega t + 30^\circ\right) A, v_1(t) = 4\sin\left(\omega t + 45^\circ\right) V, v_2(t) = 0, i_2(t) = \sin\left(\omega t + 60^\circ\right) A$$

Determine the value of $v_1(t)$ in Ampere if $i_1(t) = 0.5\sin(\omega t + 15^\circ) A$, $v_2(t) = \sin(\omega t) V$

Difficulty level   ○ Easy   ● Normal   ○ Hard
Calculation amount   ○ Small   ● Normal   ○ Large
1. $\sin(\omega t)$ V
2. $\sin(\omega t - 60^\circ)$ V
3. $0.5\sin(\omega t + 30^\circ)$ V
4. $4\sin(\omega t - 15^\circ)$ V

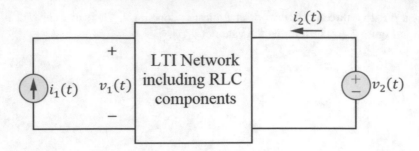

Figure 7.5 The circuit of problem 7.5

7.6. The two-port network, illustrated in Figure 7.6, is linear time-invariant (LTI). The tests below have been carried out on the resistive load. Calculate the impedance of the network.

Test I: If $R_L = \infty$, $|\mathbf{V_{out}}| = 13\ V$
Test II: If $R_L = 3\ \Omega$, $|\mathbf{V_{out}}| = 3\ V$
Test III: If $R_L = 14\ \Omega$, $|\mathbf{V_{out}}| = 9.1\ V$

Difficulty level   ○ Easy   ○ Normal   ● Hard
Calculation amount   ○ Small   ● Normal   ○ Large
1. $2\ \Omega$
2. $12\ \Omega$
3. $(12 + j2)\ \Omega$
4. $(2 + j12)\ \Omega$

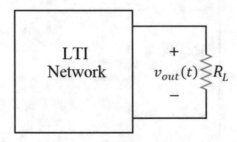

Figure 7.6 The circuit of problem 7.6

7.7. In the linear time-invariant (LTI) two-port network that only includes resistors, inductors, and capacitors, when an impulse function is applied on the left-side port, the voltage of $v_2(t) = 2\ V$ is measured on the right-side port (see Figure 7.7.1). Now, the same network is evaluated in another test, presented in Figure 7.7.2. Determine the value of $\widetilde{v}_1(t)$.

Difficulty level   ○ Easy   ○ Normal   ● Hard
Calculation amount   ○ Small   ● Normal   ○ Large
1. $2 \cos (t)\ V$
2. $\cos(t)\ V$
3. $2 \sin (t)\ V$
4. $\sin(t)\ V$

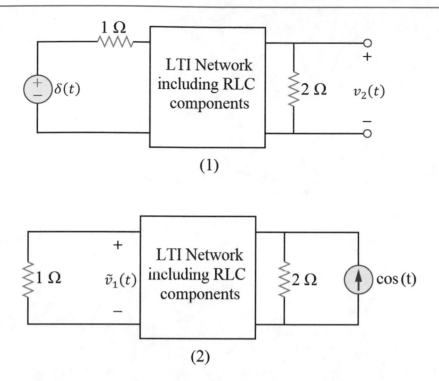

(1)

(2)

Figure 7.7 The circuit of problem 7.7

References

1. Rahmani-Andebili, M. (2020). DC Electrical circuit analysis: Practice problems, methods, and solutions, Springer Nature.
2. Rahmani-Andebili, M. (2020). AC Electrical circuit analysis: Practice problems, methods, and solutions, Springer Nature.

Solutions of Problems: Network Theorems (Tellegen's and Linear Time-Invariant Network Theorems)

8

Abstract

In this chapter, the problems of the seventh chapter are fully solved, in detail, step-by-step, and with different methods.

8.1. Based on the information given in the problem, we have [1–2]:

$$\text{Test I}: v_s = 7\,V, i_s = 3\,A, i_{out} = 1\,A \tag{1}$$

$$\text{Test II}: v_s = 9\,V, i_s = 1\,A, i_{out} = 3\,A \tag{2}$$

$$\text{Test III}: v_s = 15\,V, i_s = 9\,A, i_{out} = \text{Unknown} \tag{3}$$

Since the network only includes linear time-invariant (LTI) resistors and dependent sources, the following relation for the output current is held:

$$i_{out} = \alpha i_s + \beta v_s \tag{4}$$

Solving (1), (2), and (4):

$$\begin{cases} 1 = 3\alpha + 7\beta \\ 3 = \alpha + 9\beta \end{cases} \Rightarrow \alpha = -0.6, \beta = 0.4 \tag{5}$$

Solving (3), (4), and (5):

$$i_{out} = \alpha i_s + \beta v_s = -0.6 \times 9 + 0.4 \times 15 = 0.6\,A$$

Choice (2) is the answer.

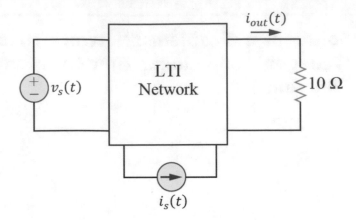

Figure 8.1 The circuit of solution of problem 8.1

8.2. Based on Tellegen's theorem, we can write the following equation for the network:

$$v_1(t)\left(-\tilde{i}_1(t)\right) + v_2(t)\tilde{i}_2(t) = \tilde{v}_1(t)(-i_1(t)) + \tilde{v}_2(t)i_2(t) \tag{1}$$

In (1), a negative sign was applied for the polarity of the current of the left-side voltage source, as its current flows toward the network.

Based on the information given in the problem, we have:

$$\text{Test I}: v_1(t) = 30t \; V, v_2(t) = 0, i_1(t) = 5t \; A, i_2(t) = 2t \; A \tag{2}$$

$$\text{Test II}: v_1(t) = 30t + 60 \; V, v_2(t) = 60t + 15 \; V, i_1(t) = \text{Unknown} \tag{3}$$

Solving (1), (2), and (3):

$$30t \times \left(-\tilde{i}_1(t)\right) + 0 \times \tilde{i}_2(t) = (30t + 60) \times (-5t) + (60t + 15) \times 2t$$

$$\Rightarrow \tilde{i}_1(t) = \frac{-150t^2 - 300t + 120t^2 + 30t}{-30t} = \frac{-30t^2 - 270t}{-30t} = (t + 9) \; A$$

Choice (1) is the answer.

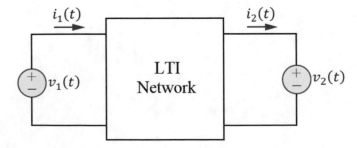

Figure 8.2 The circuit of solution of problem 8.2

8.3. Based on the information given in the problem, we have:

$$\text{Test I}: \text{If } v_1 = 3 \; V, i_s = 3 \; A, i_{out} = 6 \; A \tag{1}$$

$$\text{Test II} : v_1 = 0 \, A, i_s = -2 \, A, i_{out} = 2 \, A \tag{2}$$

$$\text{Test III} : v_1 = -2 \, V, i_s = 0 \, A, i_{out} = \text{Unknown} \tag{3}$$

Since the network is linear time-invariant (LTI), the following relation for the output current is held:

$$i_{out} = \alpha v_1 + \beta i_s \tag{4}$$

Solving (1), (2), and (4):

$$\begin{cases} 6 = 3\alpha + 3\beta \\ 2 = 0 \times \alpha + (-2)\beta \end{cases} \Rightarrow \alpha = 3, \beta = -1 \tag{5}$$

Solving (3), (4), and (5):

$$i_{out} = \alpha v_1 + \beta i_s = 3 \times (-2) + (-1) \times 0 = -6 \, A$$

Choice (4) is the answer.

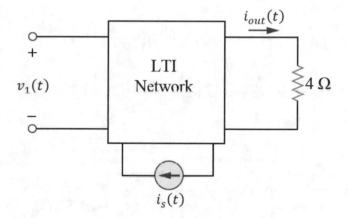

Figure 8.3 The circuit of solution of problem 8.3

8.4. Based on Tellegen's theorem, we can write:

$$v_1(t)\widetilde{i_1}(t) + v_2(t)\widetilde{i_2}(t) + v_3(t)\widetilde{i_3}(t) = \widetilde{v_1}(t)i_1(t) + \widetilde{v_2}(t)i_2(t) + \widetilde{v_3}(t)i_3(t) \tag{1}$$

From the circuits of Figs. 8.4.1–2, we have:

$$\text{Fig.1} : \begin{cases} v_1(t) = \text{Unknown} \\ v_2(t) = \left(e^{-t} - e^{-2t}\right)u(t) \, V \\ v_3(t) = 0 \\ i_1(t) = -e^{-2t}u(t) \, A \\ i_2(t) = 0 \\ i_3(t) = \left(2e^{-2t} - e^{-t}\right)u(t) \, A \end{cases} , \quad \text{Fig.2} : \begin{cases} \widetilde{v_1}(t) = 0 \\ \widetilde{v_2}(t) = \text{Unknown} \\ \widetilde{v_3}(t) = e^{-t}u(t) \, V \\ \widetilde{i_1}(t) = 0 \\ \widetilde{i_2}(t) = -e^{-t}u(t) \, A \\ \widetilde{i_3}(t) = \text{Unknown} \end{cases} \tag{2}$$

In (2), based on Tellegen's theorem, a negative sign was applied for the polarity of the currents flowing toward the network. Now, for this problem, we should transfer to Laplace domain. Therefore, the quantities are as follows:

$$\text{Fig.1}: \begin{cases} V_1(s) = \text{Unknown} \\ V_2(s) = \dfrac{1}{s+1} - \dfrac{1}{s+2} \\ V_3(s) = 0 \\ I_1(s) = -\dfrac{1}{s+2} \\ I_2(s) = 0 \\ I_3(s) = \dfrac{2}{s+2} - \dfrac{1}{s+1} \end{cases} , \quad \text{Fig.2}: \begin{cases} \widetilde{V}_1(s) = 0 \\ \widetilde{V}_2(s) = \text{Unknown} \\ \widetilde{V}_3(s) = \dfrac{1}{s+1} \\ \widetilde{I}_1(s) = 0 \\ \widetilde{I}_2(s) = -\dfrac{1}{s+1} \\ \widetilde{I}_3(s) = \text{Unknown} \end{cases} \tag{3}$$

For this problem, we should use Tellegen's theorem in Laplace domain, as follows:

$$V_1(s)\widetilde{I}_1(s) + V_2(s)\widetilde{I}_2(s) + V_3(s)\widetilde{I}_3(s) = \widetilde{V}_1(s)I_1(s) + \widetilde{V}_2(s)I_2(s) + \widetilde{V}_3(s)I_3(s) \tag{4}$$

Solving (3) and (4):

$$V_1(s) \times 0 + \left(\frac{1}{s+1} - \frac{1}{s+2}\right)\left(-\frac{1}{s+1}\right) + 0 \times \widetilde{I}_3(s) = \widetilde{V}_1(s)\left(-\frac{1}{s+2}\right) + \widetilde{V}_2(s) \times 0 + \left(\frac{1}{s+1}\right)\left(\frac{2}{s+2} - \frac{1}{s+1}\right)$$

$$\Rightarrow \widetilde{V}_1(s)\left(\frac{1}{s+2}\right) = \left(\frac{1}{s+1}\right)\left(\frac{1}{s+2}\right) \Rightarrow \widetilde{V}_1(s) = \frac{1}{s+1} \overset{L^{-1}}{\Rightarrow} v_1(t) = e^{-t}u(t) \ V$$

Choice (1) is the answer.

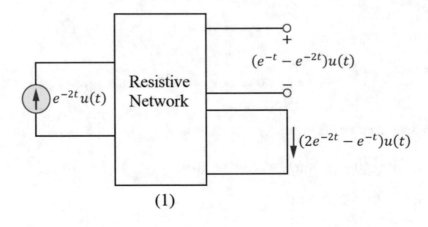

(1)

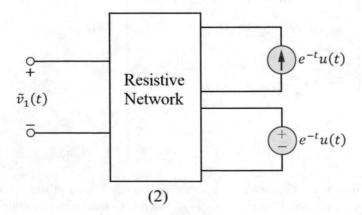

(2)

Figure 8.4 The circuit of solution of problem 8.4

8.5. Based on Tellegen's theorem, we can write the following equation for the network:

$$v_1(t)\tilde{i}_1(t) + v_2(t)\tilde{i}_2(t) = \tilde{v}_1(t)i_1(t) + \tilde{v}_2(t)i_2(t) \tag{1}$$

Based on the information given in the problem, we have:

$$
\text{Test 1}: \begin{cases} v_1(t) = 4\sin\left(\omega t + 45°\right) V \\ v_2(t) = 0 \\ i_1(t) = 2\sin\left(\omega t + 30°\right) A \\ i_2(t) = \sin\left(\omega t + 60°\right) A \end{cases}, \quad
\text{Test 2}: \begin{cases} \tilde{v}_1(t) = \text{Requested} \\ \tilde{v}_2(t) = \sin\left(\omega t\right) V \\ \tilde{i}_1(t) = 0.5\sin\left(\omega t + 15°\right) A \\ \tilde{i}_2(t) = \text{Unknown} \end{cases} \tag{2}
$$

We need to solve the problem in phasor domain. Therefore, the quantities will be as follows:

$$
\text{Test 1}: \begin{cases} \mathbf{V_1} = 4e^{j45°} \\ \mathbf{V_2} = 0 \\ \mathbf{I_1} = 2e^{j30°} \\ \mathbf{I_2} = e^{j60°} \end{cases}, \quad
\text{Test 2}: \begin{cases} \tilde{\mathbf{V}}_1 = \text{Requested} \\ \tilde{\mathbf{V}}_2 = 1 \\ \tilde{\mathbf{I}}_1 = 0.5e^{j15°} \\ \tilde{\mathbf{I}}_2 = \text{Unknown} \end{cases} \tag{3}
$$

Tellegen's theorem in phasor domain for this problem is as follows:

$$\mathbf{V_1}\tilde{\mathbf{I}}_1 + \mathbf{V_2}\tilde{\mathbf{I}}_2 = \tilde{\mathbf{V}}_1\mathbf{I_1} + \tilde{\mathbf{V}}_2\mathbf{I_2} \tag{4}$$

Solving (3) and (4):

$$4e^{j45°} \times 0.5e^{j15°} + 0 \times \tilde{\mathbf{I}}_2 = \tilde{\mathbf{V}}_1 \times 2e^{j30°} + 1 \times e^{j60°} \Rightarrow \tilde{\mathbf{V}}_1 \times 2e^{j30°} = e^{j60°} \Rightarrow \tilde{\mathbf{V}}_1 = 0.5e^{j30°}$$

By transferring to time domain, we have:

$$\tilde{v}_1(t) = 0.5\sin\left(\omega t + 30°\right) V$$

Choice (3) is the answer.

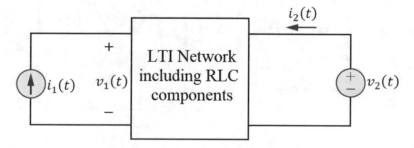

Figure 8.5 The circuit of solution of problem 8.5

8.6. Based on the information given in the problem, we have:

$$\text{Test I} : \text{If } R_L = \infty, |\mathbf{V_{out}}| = 13 \ V \tag{1}$$

$$\text{Test II} : \text{If } R_L = 3 \ \Omega, |\mathbf{V_{out}}| = 3 \ V \tag{2}$$

$$\text{Test III} : \text{If } R_L = 14 \ \Omega, |\mathbf{V_{out}}| = 9.1 \ V \tag{3}$$

Since the network is linear time-invariant (LTI), it can be replaced by its Thevenin equivalent circuit, as can be seen in Figure 8.6.2.

By using Figure 8.6.2 and (1), we can conclude:

$$|\mathbf{V_{Th}}| = |\mathbf{V_{out}}| = 13 \ V \tag{4}$$

Using Figure 8.6.2 and applying voltage division and (2) and (4):

$$|\mathbf{V_{out}}| = \left|\frac{R_L}{R_L + a + jb} \times \mathbf{V_{Th}}\right| \Rightarrow 3 = \left|\frac{3}{3 + a + jb} \times 13\right| \Rightarrow 1 = \left|\frac{13}{3 + a + jb}\right| \Rightarrow |3 + a + jb| = 13$$

$$\Rightarrow (3 + a)^2 + b^2 = 169 \tag{5}$$

Using Figure 8.6.2 and applying voltage division and (3) and (4):

$$|\mathbf{V_{out}}| = \left|\frac{R_L}{R_L + a + jb} \times \mathbf{V_{Th}}\right| \Rightarrow 9.1 = \left|\frac{14}{14 + a + jb} \times 13\right| \Rightarrow 9.1 = \left|\frac{182}{14 + a + jb}\right| \Rightarrow |14 + a + jb| = 20$$

$$\Rightarrow (14 + a)^2 + b^2 = 400 \tag{6}$$

Solving (5) and (6):

$$a = 2, b = 12 \Rightarrow Z_{Th} = (2 + j12) \ \Omega$$

Choice (4) is the answer.

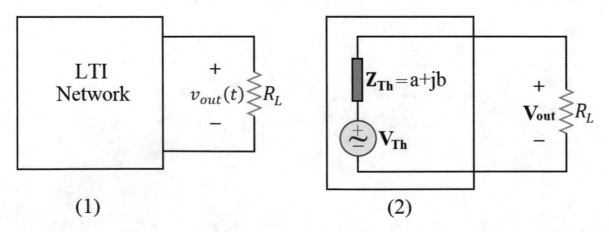

Figure 8.6 The circuit of solution of problem 8.6

8.7. First, we should define a new network with the boundary shown in Figs. 8.7.3–4. Then, based on Tellegen's theorem, we can write:

$$v_1(t)\widetilde{i}_1(t) + v_2(t)\widetilde{i}_2(t) = \widetilde{v}_1(t)i_1(t) + \widetilde{v}_2(t)i_2(t) \tag{1}$$

From the circuits of Figs. 8.7.3–4 as well as based on the information given in the problem, we have:

$$\text{Fig.3}: \begin{cases} v_1(t) = \delta(t) \\ v_2(t) = 2\ V \\ i_1(t) = \text{Unknown} \\ i_2(t) = 0 \end{cases}, \quad \text{Fig.4}: \begin{cases} \widetilde{v}_1(t) = 0 \\ \widetilde{v}_2(t) = \text{Unknown} \\ \widetilde{i}_1(t) = \text{Unknown} \\ \widetilde{i}_2(t) = -\cos(t)\ A \end{cases} \tag{2}$$

In (2), based on Tellegen's theorem, a negative sign was applied for the polarity of the current flowing toward the network. Now, we need to solve the problem in phasor domain. Therefore, the quantities will be as follows:

$$\text{Fig.3}: \begin{cases} \mathbf{V}_1 = 1 \\ \mathbf{V}_2 = 2 \\ \mathbf{I}_1 = \text{Unknown} \\ \mathbf{I}_2 = 0 \end{cases}, \quad \text{Fig.4}: \begin{cases} \widetilde{\mathbf{V}}_1 = 0 \\ \widetilde{\mathbf{V}}_2 = \text{Unknown} \\ \widetilde{\mathbf{I}}_1 = \text{Unknown} \\ \widetilde{\mathbf{I}}_2 = -1e^{j0} = -1 \end{cases} \tag{3}$$

Tellegen's theorem in phasor domain for this problem is as follows:

$$\mathbf{V}_1\widetilde{\mathbf{I}}_1 + \mathbf{V}_2\widetilde{\mathbf{I}}_2 = \widetilde{\mathbf{V}}_1\mathbf{I}_1 + \widetilde{\mathbf{V}}_2\mathbf{I}_2 \tag{4}$$

Solving (3) and (4):

$$1 \times \widetilde{\mathbf{I}}_1 + 2 \times (-1) = 0 \times \mathbf{I}_1 + \widetilde{\mathbf{V}}_2 \times 0 \Rightarrow \widetilde{\mathbf{I}}_1 = 2 \tag{5}$$

By transferring to time domain, we have:

$$\widetilde{i}_1(t) = 2\cos(t)\ A \tag{6}$$

Using (6) and Ohm's law in Figure 8.7.4:

$$\widetilde{v}_1(t) = 1 \times \widetilde{i}_1(t) = 2\ \cos(t)\ V$$

Choice (1) is the answer.

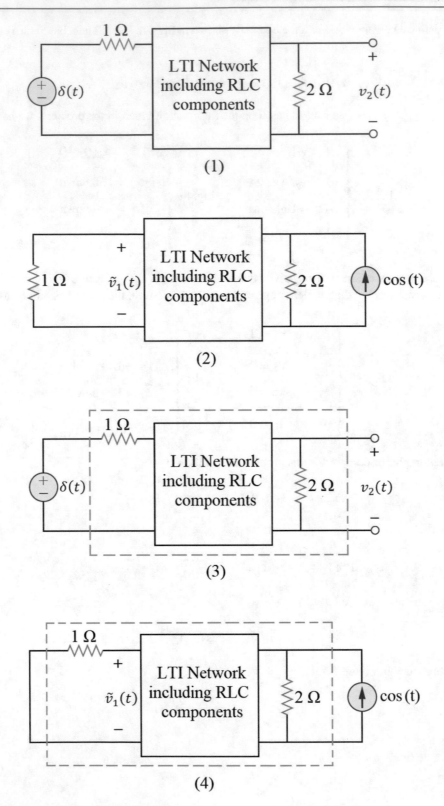

Figure 8.7 The circuit of solution of problem 8.7

References

1. Rahmani-Andebili, M. (2020). DC Electrical circuit analysis: Practice problems, methods, and solutions, Springer Nature.
2. Rahmani-Andebili, M. (2020). AC Electrical circuit analysis: Practice problems, methods, and solutions, Springer Nature.

Problems: Two-Port Networks

9

Abstract

In this chapter, the basic and advanced problems concerned with the determination of different matrices of a two-port network, that is, impedance matrix, admittance matrix, hybrid matrix, and transmission matrix as well as the series and parallel connection of the two-port networks, are presented. In this chapter, the problems are categorized in different levels based on their difficulty levels (easy, normal, and hard) and calculation amounts (small, normal, and large). Additionally, the problems are ordered from the easiest problem with the smallest computations to the most difficult problems with the largest calculations.

9.1. For the two-port network, shown in Figure 9.1, determine the impedance matrix ([Z]) [1–2].

Difficulty level   ● Easy   ○ Normal   ○ Hard
Calculation amount   ● Small   ○ Normal   ○ Large

1. $\begin{bmatrix} 1 & 1 \\ 1 & 1 \end{bmatrix}$

2. $\begin{bmatrix} -1 & 1 \\ 1 & -1 \end{bmatrix}$

3. $\begin{bmatrix} 1 & -1 \\ -1 & 1 \end{bmatrix}$

4. $\begin{bmatrix} -1 & -1 \\ -1 & -1 \end{bmatrix}$

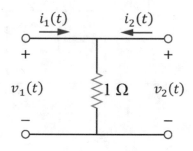

Figure 9.1 The circuit of problem 9.1

9.2. For the circuit, shown in Figure 9.2, determine the transmission matrix ([T]).

    Difficulty level   ● Easy   ○ Normal   ○ Hard
    Calculation amount ● Small   ○ Normal   ○ Large

1. $\begin{bmatrix} -1 & 0 \\ -1 & -1 \end{bmatrix}$

2. $\begin{bmatrix} -1 & 0 \\ 1 & -1 \end{bmatrix}$

3. $\begin{bmatrix} 0 & 1 \\ 1 & 0 \end{bmatrix}$

4. $\begin{bmatrix} 1 & 0 \\ 1 & 1 \end{bmatrix}$

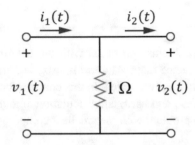

Figure 9.2 The circuit of problem 9.2

9.3. In the circuit of Figure 9.3, determine the hybrid parameter of h_{21}.

    Difficulty level   ○ Easy   ● Normal   ○ Hard
    Calculation amount ● Small   ○ Normal   ○ Large

1. $-\frac{\alpha + j\omega R_2 C}{1 + j\omega R_2 C}$

2. $\frac{\alpha + j\omega R_2}{1 + j\omega R_2}$

3. $\frac{1 + j\omega R_2 C}{1 + j\omega}$

4. None of them

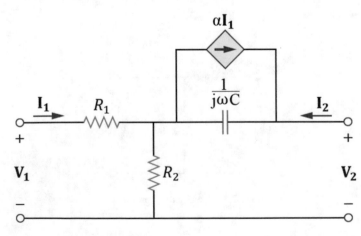

Figure 9.3 The circuit of problem 9.3

9.4. For the two-port network, shown in Figure 9.4, determine the admittance matrix ($[Y]$).

1. $\begin{bmatrix} 2(s+1) & -(s+1) \\ -(s+1) & 2(s+1) \end{bmatrix}$

2. $\begin{bmatrix} \dfrac{2}{s}+2 & -\left(\dfrac{1}{s}+1\right) \\ -\left(\dfrac{1}{s}+1\right) & \dfrac{2}{s}+2 \end{bmatrix}$

3. $\begin{bmatrix} 2\dfrac{s+1}{s} & -(s+1) \\ -(s+1) & 2\dfrac{s+1}{s} \end{bmatrix}$

4. None of them

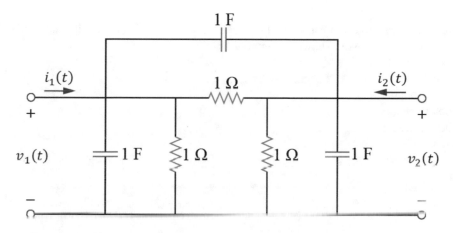

Figure 9.4 The circuit of problem 9.4

9.5. For the two-port network, shown in Figure 9.5, determine the impedance matrix ($[Z]$).

1. $\begin{bmatrix} \alpha+R & R-j\dfrac{\beta}{C\omega} \\ \alpha+R & R+j\dfrac{\beta}{C\omega} \end{bmatrix}$

2. $\begin{bmatrix} \alpha+R & j\dfrac{\beta}{C\omega} \\ \alpha & R+j\dfrac{\beta}{C\omega} \end{bmatrix}$

3. $\begin{bmatrix} \alpha+R & \alpha+R \\ \alpha+R+j\dfrac{\beta}{C\omega} & \alpha+R-j\dfrac{1-\beta}{C\omega} \end{bmatrix}$

4. None of them

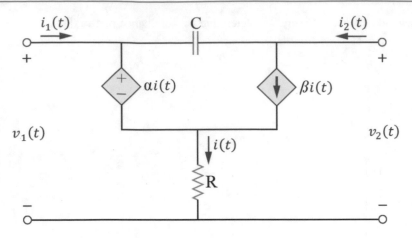

Figure 9.5 The circuit of problem 9.5

9.6. Which one of the following choices is true about the impedance matrix ($[Z]$), the admittance matrix ($[Y]$), and the hybrid matrix ($[H]$) of the two-port network shown in Figure 9.6?

Difficulty level   ○ Easy   ● Normal   ○ Hard
Calculation amount   ○ Small   ● Normal   ○ Large

1. All of them are available.
2. Only $[Z]$ is not available.
3. Only $[Y]$ is not available.
4. Only $[H]$ is not available.

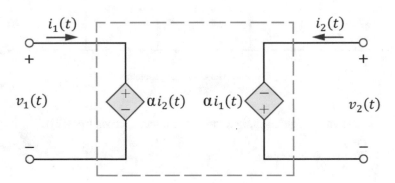

Figure 9.6 The circuit of problem 9.6

9.7. For the circuit, shown in Figure 9.7, determine the admittance matrix ($[Y]$).

Difficulty level   ○ Easy   ● Normal   ○ Hard
Calculation amount   ○ Small   ● Normal   ○ Large

1. $\begin{bmatrix} 4 & -1 \\ -3 & 2 \end{bmatrix}$

2. $\begin{bmatrix} 2 & -1 \\ -3 & 4 \end{bmatrix}$

3. $\begin{bmatrix} 4 & -3 \\ -1 & 2 \end{bmatrix}$

4. None of them

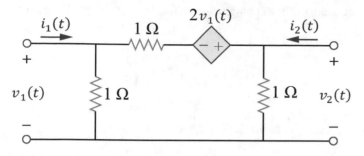

Figure 9.7 The circuit of problem 9.7

9.8. For what value of α the two-port network, shown in Figure 9.8, does not have the admittance matrix ($[Y]$)?

Difficulty level   ○ Easy   ● Normal   ○ Hard
Calculation amount   ○ Small   ● Normal   ○ Large
1. $\frac{11}{6}$
2. $-\frac{11}{6}$
3. $\frac{6}{11}$
4. $-\frac{6}{11}$

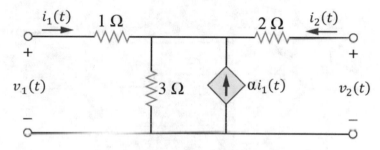

Figure 9.8 The circuit of problem 9.8

9.9. For what value of α and β the two-port network, shown in Figure 9.9, has the admittance matrix in the following form?

$$[Y] = \begin{bmatrix} 2 & -2 \\ 0 & -2 \end{bmatrix}$$

Difficulty level   ○ Easy   ● Normal   ○ Hard
Calculation amount   ○ Small   ● Normal   ○ Large
1. $\alpha = 1, \beta = -1$
2. $\alpha = 1, \beta = 1$
3. $\alpha = -1, \beta = -1$
4. $\alpha = -1, \beta = 1$

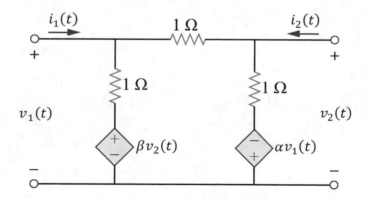

Figure 9.9 The circuit of problem 9.9

9.10. Determine the admittance matrix of the configuration shown in Figure 9.10. The admittance matrix of each network is as follows:

$$[Y_1] = \begin{bmatrix} 1 & 2 \\ 2 & 3 \end{bmatrix}, [Y_2] = \begin{bmatrix} 1 & 2 \\ 2 & 6 \end{bmatrix}$$

Difficulty level   ○ Easy   ○ Normal   ● Hard
Calculation amount   ● Small   ○ Normal   ○ Large

1. $\begin{bmatrix} 2 & 4 \\ 4 & 9 \end{bmatrix}$

2. $\begin{bmatrix} 0 & 0 \\ 0 & 3 \end{bmatrix}$

3. $\begin{bmatrix} 1 & 1 \\ 1 & 1 \end{bmatrix}$

4. None of them

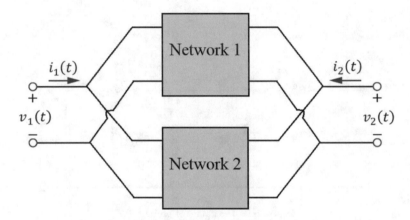

Figure 9.10 The circuit of problem 9.10

9.11. Which one of the choices presents the correct relation between the input and output voltages and currents for the configuration of Figure 9.11 if the impedance matrices of the networks are as follows?

$$[Z_1] = \begin{bmatrix} 1 & 2 \\ 2 & 3 \end{bmatrix}, [Z_2] = \begin{bmatrix} 5 & 6 \\ 6 & 7 \end{bmatrix}$$

Difficulty level   ○ Easy   ○ Normal   ● Hard
Calculation amount   ● Small   ○ Normal   ○ Large

1. $\begin{bmatrix} v_1 \\ v_2 \end{bmatrix} = \begin{bmatrix} 10 & 8 \\ 8 & 5 \end{bmatrix} \begin{bmatrix} i_1 \\ i_2 \end{bmatrix}$

2. $\begin{bmatrix} v_1 \\ v_2 \end{bmatrix} = \begin{bmatrix} 4 & 8 \\ 8 & 6 \end{bmatrix} \begin{bmatrix} i_1 \\ i_2 \end{bmatrix}$

3. $\begin{bmatrix} v_1 \\ v_2 \end{bmatrix} = \begin{bmatrix} 8 & 6 \\ 6 & 10 \end{bmatrix} \begin{bmatrix} i_1 \\ i_2 \end{bmatrix}$

4. $\begin{bmatrix} v_1 \\ v_2 \end{bmatrix} = \begin{bmatrix} 6 & 8 \\ 8 & 10 \end{bmatrix} \begin{bmatrix} i_1 \\ i_2 \end{bmatrix}$

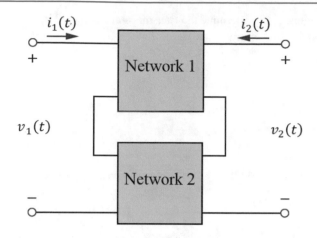

Figure 9.11 The circuit of problem 9.11

9.12. For the two-port network, shown in Figure 9.12, determine the hybrid matrix ([H]).

Difficulty level ○ Easy ○ Normal ● Hard
Calculation amount ○ Small ○ Normal ● Large

1. $\begin{bmatrix} 0 & \dfrac{1+\beta_1}{1+\beta_2} \\[3mm] -\dfrac{1+\alpha_1}{1+\alpha_2} & 0 \end{bmatrix}$

2. $\begin{bmatrix} 0 & \dfrac{1+\beta_2}{1+\beta_1} \\[3mm] -\dfrac{1+\alpha_2}{1+\alpha_1} & 0 \end{bmatrix}$

3. $\begin{bmatrix} 0 & \dfrac{1+\alpha_1}{1+\alpha_2} \\[3mm] -\dfrac{1+\beta_1}{1+\beta_2} & 0 \end{bmatrix}$

4. $\begin{bmatrix} 0 & \dfrac{1+\alpha_2}{1+\alpha_1} \\[3mm] -\dfrac{1+\beta_2}{1+\beta_1} & 0 \end{bmatrix}$

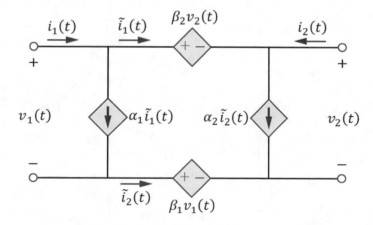

Figure 9.12 The circuit of problem 9.12

9.13. For the circuit, shown in Figure 9.13, determine the transmission matrix ($[T]$).

Difficulty level   ○ Easy   ○ Normal   ● Hard
Calculation amount   ○ Small   ○ Normal   ● Large

1. $\begin{bmatrix} 26 & 15 \\ 7 & 6 \end{bmatrix}$

2. $\begin{bmatrix} 26 & 15 \\ 40 & 20 \end{bmatrix}$

3. $\begin{bmatrix} 26 & 15 \\ 21 & 15 \end{bmatrix}$

4. $\begin{bmatrix} 26 & 15 \\ 45 & 26 \end{bmatrix}$

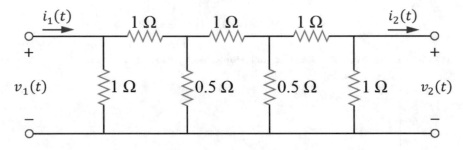

Figure 9.13 The circuit of problem 9.13

References

1. Rahmani-Andebili, M. (2020). DC Electrical circuit analysis: Practice problems, methods, and solutions, Springer Nature.
2. Rahmani-Andebili, M. (2020). AC Electrical circuit analysis: Practice problems, methods, and solutions, Springer Nature.

Abstract

In this chapter, the problems of the ninth chapter are fully solved, in detail, step-by-step, and with different methods.

10.1. The impedance matrix ([Z]) of a network in time domain is as follows [1–2]:

$$\begin{bmatrix} v_1(t) \\ v_2(t) \end{bmatrix} = \begin{bmatrix} z_{11} & z_{12} \\ z_{21} & z_{22} \end{bmatrix} \begin{bmatrix} i_1(t) \\ i_2(t) \end{bmatrix} \Rightarrow [Z] = \begin{bmatrix} z_{11} & z_{12} \\ z_{21} & z_{22} \end{bmatrix} \tag{1}$$

Applying KVL in the left-side mesh:

$$-v_1(t) + 1 \times (i_1(t) + i_2(t)) = 0 \Rightarrow v_1(t) = i_1(t) + i_2(t) \tag{2}$$

Applying KVL in the right-side mesh:

$$-v_2(t) + 1 \times (i_1(t) + i_2(t)) = 0 \Rightarrow v_2(t) = i_1(t) + i_2(t) \tag{3}$$

Solving (1), (2), and (3):

$$\begin{bmatrix} v_1(t) \\ v_2(t) \end{bmatrix} = \begin{bmatrix} 1 & 1 \\ 1 & 1 \end{bmatrix} \begin{bmatrix} i_1(t) \\ i_2(t) \end{bmatrix} \Rightarrow [Z] = \begin{bmatrix} 1 & 1 \\ 1 & 1 \end{bmatrix}$$

Choice (1) is the answer.

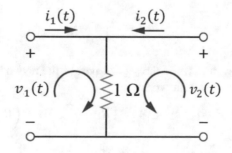

Figure 10.1 The circuit of solution of problem 10.1

10.2. The transmission matrix ($[T]$) of a network is defined as follows:

$$\begin{bmatrix} v_1(t) \\ i_1(t) \end{bmatrix} = \begin{bmatrix} t_{11} & t_{12} \\ t_{21} & t_{22} \end{bmatrix} \begin{bmatrix} v_2(t) \\ i_2(t) \end{bmatrix} \Rightarrow [T] = \begin{bmatrix} t_{11} & t_{12} \\ t_{21} & t_{22} \end{bmatrix} \tag{1}$$

Applying KVL in the loop:

$$-v_1(t) + v_2(t) = 0 \Rightarrow v_1(t) = v_2(t) \tag{2}$$

Applying KVL in the right-side mesh:

$$-v_2(t) + 1 \times (i_1(t) - i_2(t)) = 0 \Rightarrow i_1(t) = v_2(t) + i_2(t) \tag{3}$$

Solving (1), (2), and (3):

$$\begin{bmatrix} v_1(t) \\ i_1(t) \end{bmatrix} = \begin{bmatrix} 1 & 0 \\ 1 & 1 \end{bmatrix} \begin{bmatrix} v_2(t) \\ i_2(t) \end{bmatrix} \Rightarrow [T] = \begin{bmatrix} 1 & 0 \\ 1 & 1 \end{bmatrix}$$

Choice (4) is the answer.

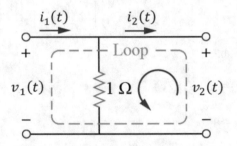

Figure 10.2 The circuit of solution of problem 10.2

10.3. Hybrid matrix ($[H]$) in phasor domain is defined as follows:

$$\begin{bmatrix} \mathbf{V_1} \\ \mathbf{I_2} \end{bmatrix} = \begin{bmatrix} h_{11} & h_{12} \\ h_{21} & h_{22} \end{bmatrix} \begin{bmatrix} \mathbf{I_1} \\ \mathbf{V_2} \end{bmatrix} \Rightarrow [H] = \begin{bmatrix} h_{11} & h_{12} \\ h_{21} & h_{22} \end{bmatrix} \tag{1}$$

The hybrid parameter of h_{21} can be determined as follows:

$$h_{21} = \left. \frac{\mathbf{I_2}}{\mathbf{I_1}} \right|_{\mathbf{V_2}=0} \tag{2}$$

Figure 10.3.2 illustrates the main circuit, while the second port of the network has been short-circuited ($\mathbf{V_2} = 0$). By applying KVL in the right-side mesh, we can write:

$$\frac{1}{j\omega C}(\mathbf{I_2} + \alpha \mathbf{I_1}) + R_2(\mathbf{I_1} + \mathbf{I_2}) = 0 \Rightarrow \left(R_2 + \frac{\alpha}{j\omega C} \right) \mathbf{I_1} + \left(R_2 + \frac{1}{j\omega C} \right) \mathbf{I_2} = 0$$

$$\Rightarrow h_{21} = \left. \frac{\mathbf{I_2}}{\mathbf{I_1}} \right|_{\mathbf{V_2}=0} = -\frac{R_2 + \frac{\alpha}{j\omega C}}{R_2 + \frac{1}{j\omega C}} = -\frac{\alpha + j\omega C R_2}{1 + j\omega C R_2}$$

Choice (1) is the answer.

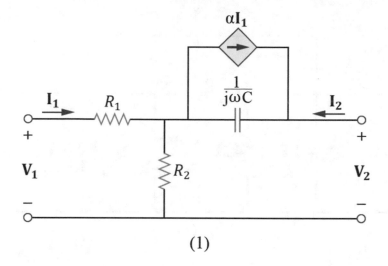

(1)

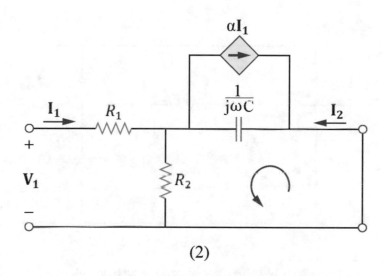

(2)

Figure 10.3 The circuit of solution of problem 10.3

10.4. The circuit of Figure 10.4.2 shows the main circuit in Laplace domain. The impedances of the components are as follows:

$$Z_R = R \Rightarrow Z_{1\,\Omega} = 1\,\Omega$$

$$Z_C = \frac{1}{Cs} \Rightarrow Z_{1\,F} = \frac{1}{s}\,\Omega$$

For this two-port network, the admittance matrix ($[Y]$) is the same as the nodal admittance matrix ($[Y_{nodal}]$). The circuit includes two supernodes shown in Figure 10.4.2. The nodal admittance matrix can be determined as follows:

$$[Y_{nodal}] = \begin{bmatrix} \sum\limits_{j=1} y_{1j} & -y_{12} \\ -y_{21} & \sum\limits_{j=1} y_{2j} \end{bmatrix} = \begin{bmatrix} s+1+s+1 & -(s+1) \\ -(s+1) & s+1+s+1 \end{bmatrix} = \begin{bmatrix} 2(s+1) & -(s+1) \\ -(s+1) & 2(s+1) \end{bmatrix}$$

$$[Y] = [Y_{nodal}] = \begin{bmatrix} 2(s+1) & -(s+1) \\ -(s+1) & 2(s+1) \end{bmatrix}$$

Choice (1) is the answer.

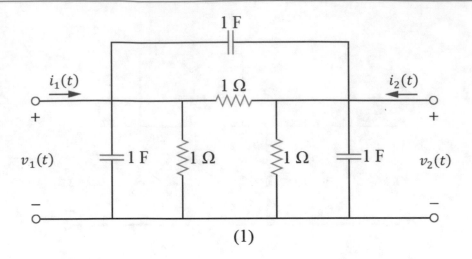

(1)

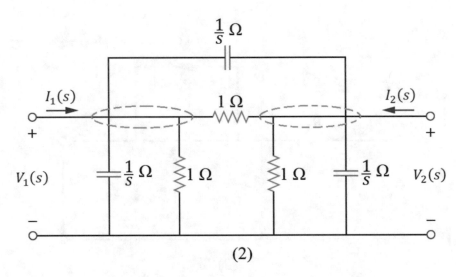

(2)

Figure 10.4 The circuit of solution of problem 10.4

10.5. Impedance matrix ([Z]) in phasor domain is defined as follows:

$$\begin{bmatrix} \mathbf{V_1} \\ \mathbf{V_2} \end{bmatrix} = \begin{bmatrix} z_{11} & z_{12} \\ z_{21} & z_{22} \end{bmatrix} \begin{bmatrix} \mathbf{I_1} \\ \mathbf{I_2} \end{bmatrix} \Rightarrow [Z] = \begin{bmatrix} z_{11} & z_{12} \\ z_{21} & z_{22} \end{bmatrix} \tag{1}$$

Figure 10.5.2 shows the main circuit in phasor domain. By applying KCL in the cut-set, we can write:

$$\mathbf{I} = \mathbf{I_1} + \mathbf{I_2} \tag{2}$$

Applying KVL in the left-side mesh:

$$\mathbf{V_1} = \alpha \mathbf{I} + R\mathbf{I} \xrightarrow{Using\ (2)} \mathbf{V_1} = (\alpha + R)\mathbf{I_1} + (\alpha + R)\mathbf{I_2} \tag{3}$$

Applying KVL in the indicated loop:

$$\mathbf{V_2} = \frac{1}{j\omega C}(\mathbf{I_2} - \beta \mathbf{I}) + \alpha \mathbf{I} + R\mathbf{I} \xrightarrow{Using\ (2)} \mathbf{V_2} = \left(\alpha + R - \frac{\beta}{j\omega C}\right)\mathbf{I_1} + \left(\alpha + R + \frac{1-\beta}{j\omega C}\right)\mathbf{I_2} \tag{4}$$

Combining (3) and (4) in the matrix form:

$$\begin{bmatrix} V_1 \\ V_2 \end{bmatrix} = \begin{bmatrix} \alpha + R & \alpha + R \\ \alpha + R + j\dfrac{\beta}{\omega C} & \alpha + R - j\dfrac{1-\beta}{\omega C} \end{bmatrix} \begin{bmatrix} I_1 \\ I_2 \end{bmatrix} \Rightarrow [Z] = \begin{bmatrix} \alpha + R & \alpha + R \\ \alpha + R + j\dfrac{\beta}{\omega C} & \alpha + R - j\dfrac{1-\beta}{\omega C} \end{bmatrix}$$

Choice (3) is the answer.

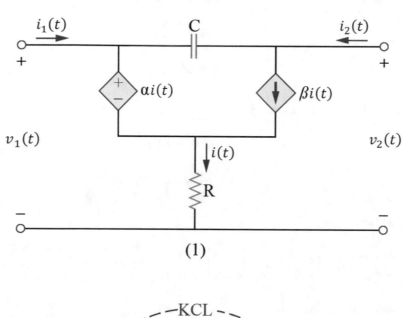

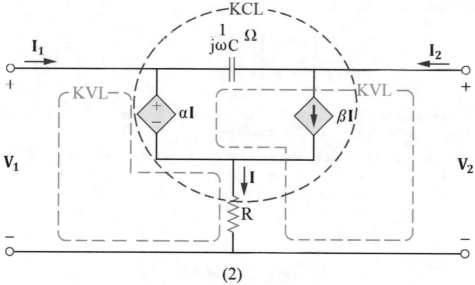

Figure 10.5 The circuit of solution of problem 10.5

10.6. The impedance matrix ([Z]), the admittance matrix ([Y]), and the hybrid matrix ([H]) in time domain are defined in the following forms:

$$\begin{bmatrix} v_1(t) \\ v_2(t) \end{bmatrix} = \begin{bmatrix} z_{11} & z_{12} \\ z_{21} & z_{22} \end{bmatrix} \begin{bmatrix} i_1(t) \\ i_2(t) \end{bmatrix} \Rightarrow [Z] = \begin{bmatrix} z_{11} & z_{12} \\ z_{21} & z_{22} \end{bmatrix} \tag{1}$$

$$\begin{bmatrix} i_1(t) \\ i_2(t) \end{bmatrix} = \begin{bmatrix} y_{11} & y_{12} \\ y_{21} & y_{22} \end{bmatrix} \begin{bmatrix} v_1(t) \\ v_2(t) \end{bmatrix} \Rightarrow [Y] = \begin{bmatrix} y_{11} & y_{12} \\ y_{21} & y_{22} \end{bmatrix} \tag{2}$$

$$\begin{bmatrix} v_1(t) \\ i_2(t) \end{bmatrix} = \begin{bmatrix} h_{11} & h_{12} \\ h_{21} & h_{22} \end{bmatrix} \begin{bmatrix} i_1(t) \\ v_2(t) \end{bmatrix} \Rightarrow [H] = \begin{bmatrix} h_{11} & h_{12} \\ h_{21} & h_{22} \end{bmatrix} \tag{3}$$

From the left-side mesh of the circuit of Figure 10.6, it is clear that:

$$v_1(t) = \alpha i_2(t) \tag{4}$$

$$\Rightarrow i_2(t) = \frac{1}{\alpha} v_1(t) \tag{5}$$

Likewise, from the right-side mesh of the circuit of Figure 10.6, it is seen that:

$$v_2(t) = -\alpha i_1(t) \tag{6}$$

$$\Rightarrow i_1(t) = -\frac{1}{\alpha} v_2(t) \tag{7}$$

Solving (1), (4), and (6):

$$\begin{bmatrix} v_1(t) \\ v_2(t) \end{bmatrix} = \begin{bmatrix} 0 & \alpha \\ -\alpha & 0 \end{bmatrix} \begin{bmatrix} i_1(t) \\ i_2(t) \end{bmatrix} \Rightarrow [Z] = \begin{bmatrix} 0 & \alpha \\ -\alpha & 0 \end{bmatrix} \tag{8}$$

Solving (2), (5), and (7):

$$\begin{bmatrix} i_1(t) \\ i_2(t) \end{bmatrix} = \begin{bmatrix} 0 & -\dfrac{1}{\alpha} \\ \dfrac{1}{\alpha} & 0 \end{bmatrix} \begin{bmatrix} v_1(t) \\ v_2(t) \end{bmatrix} \Rightarrow [Y] = \begin{bmatrix} 0 & -\dfrac{1}{\alpha} \\ \dfrac{1}{\alpha} & 0 \end{bmatrix} \tag{9}$$

However, as can be seen from (3), (4), and (5), it is impossible to form the hybrid matrix for the given network. Therefore, the hybrid matrix of the network is not available.

Consequently, [Z] and [Y] are available but [H] is not available. Choice (4) is the answer.

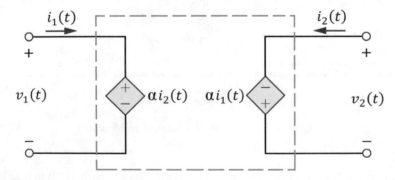

Figure 10.6 The circuit of solution of problem 10.6

10.7. The admittance matrix ($[Y]$) in time domain is defined as follows:

$$\begin{bmatrix} i_1(t) \\ i_2(t) \end{bmatrix} = \begin{bmatrix} y_{11} & y_{12} \\ y_{21} & y_{22} \end{bmatrix} \begin{bmatrix} v_1(t) \\ v_2(t) \end{bmatrix} \Rightarrow [Y] = \begin{bmatrix} y_{11} & y_{12} \\ y_{21} & y_{22} \end{bmatrix} \tag{1}$$

KCL in the left-side node:

$$-i_1(t) + \frac{v_1(t)}{1} + \frac{v_1(t) - (v_2(t) - 2v_1(t))}{1} = 0 \Rightarrow i_1(t) = 4v_1(t) - v_2(t) \tag{2}$$

KCL in the right-side node:

$$-i_2(t) + \frac{v_2(t)}{1} + \frac{v_2(t) - (v_1(t) + 2v_1(t))}{1} = 0 \Rightarrow i_2(t) = -3v_1(t) + 2v_2(t) \tag{3}$$

Solving (1), (2), and (3):

$$\begin{bmatrix} i_1(t) \\ i_2(t) \end{bmatrix} = \begin{bmatrix} 4 & -1 \\ -3 & 2 \end{bmatrix} \begin{bmatrix} v_1(t) \\ v_2(t) \end{bmatrix} \Rightarrow [Y] = \begin{vmatrix} 4 & -1 \\ -3 & 2 \end{vmatrix}$$

Choice (1) is the answer.

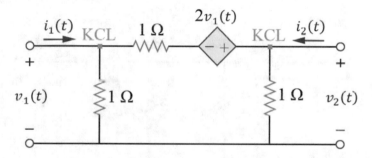

Figure 10.7 The circuit of solution of problem 10.7

10.8. The admittance matrix ($[Y]$) of a circuit is not available if the determinant of the inverse matrix of the admittance matrix ($[Y]^{-1}$) or impedance matrix (($[Z]$)) is zero. In other words:

$$\det([Z]) = 0 \tag{1}$$

The impedance matrix ($[Z]$) in time domain is as follows:

$$\begin{bmatrix} v_1(t) \\ v_2(t) \end{bmatrix} = \begin{bmatrix} z_{11} & z_{12} \\ z_{21} & z_{22} \end{bmatrix} \begin{bmatrix} i_1(t) \\ i_2(t) \end{bmatrix} \Rightarrow [Z] = \begin{bmatrix} z_{11} & z_{12} \\ z_{21} & z_{22} \end{bmatrix} \tag{2}$$

Applying KVL in the left-side mesh of the circuit of Figure 10.8.2:

$$-v_1(t) + 1 \times i_1(t) + 3((1+a)i_1(t) + i_2(t)) = 0 \Rightarrow v_1(t) = (4+3a)i_1(t) + 3i_2(t) \tag{3}$$

Applying KVL in the right-side loop of the circuit of Figure 10.8.2:

$$-v_2(t) + 2 \times i_2(t) + 3((1+a)i_1(t) + i_2(t)) = 0 \Rightarrow v_2(t) = (3+3a)i_1(t) + 5i_2(t) \tag{4}$$

Solving (2), (3), and (4):

$$\begin{bmatrix} v_1(t) \\ v_2(t) \end{bmatrix} = \begin{bmatrix} 4+3\alpha & 3 \\ 3+3\alpha & 5 \end{bmatrix} \begin{bmatrix} i_1(t) \\ i_2(t) \end{bmatrix} \Rightarrow [Z] = \begin{bmatrix} 4+3\alpha & 3 \\ 3+3\alpha & 5 \end{bmatrix} \tag{5}$$

Solving (1) and (5):

$$\det\left(\begin{bmatrix} 4+3\alpha & 3 \\ 3+3\alpha & 5 \end{bmatrix}\right) = 0 \Rightarrow (4+3\alpha)(5) - (3)(3+3\alpha) = 0 \Rightarrow 6\alpha + 11 = 0 \Rightarrow \alpha = -\frac{11}{6}$$

Choice (2) is the answer.

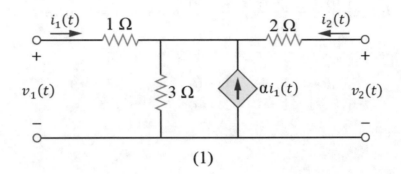

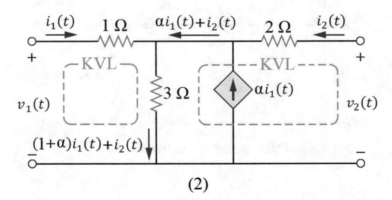

Figure 10.8 The circuit of solution of problem 10.8

10.9. Based on the information given in the problem, the admittance matrix of the circuit is as follows:

$$[Y] = \begin{bmatrix} 2 & -2 \\ 0 & -2 \end{bmatrix} \tag{1}$$

The admittance matrix ($[Y]$) of a circuit is defined as follows:

$$\begin{bmatrix} i_1(t) \\ i_2(t) \end{bmatrix} = \begin{bmatrix} y_{11} & y_{12} \\ y_{21} & y_{22} \end{bmatrix} \begin{bmatrix} v_1(t) \\ v_2(t) \end{bmatrix} \Rightarrow [Y] = \begin{bmatrix} y_{11} & y_{12} \\ y_{21} & y_{22} \end{bmatrix} \tag{2}$$

Applying KCL in the left-side node:

$$-i_1(t) + \frac{v_1(t) - \beta v_2(t)}{1} + \frac{v_1(t) - v_2(t)}{1} = 0 \Rightarrow i_1(t) = 2v_1(t) - (1+\beta)v_2(t) \tag{3}$$

Applying KCL in the right-side node:

$$-i_2(t) + \frac{v_2(t) - (-\alpha v_1(t))}{1} + \frac{v_2(t) - v_1(t)}{1} = 0 \Rightarrow i_2(t) = (\alpha - 1)v_1(t) + 2v_2(t) \tag{4}$$

Solving (2), (3), and (4):

$$\begin{bmatrix} i_1(t) \\ i_2(t) \end{bmatrix} = \begin{bmatrix} 2 & -(1+\beta) \\ \alpha - 1 & 2 \end{bmatrix} \begin{bmatrix} v_1(t) \\ v_2(t) \end{bmatrix} \Rightarrow [Y] = \begin{bmatrix} 2 & -(1+\beta) \\ \alpha - 1 & 2 \end{bmatrix} \tag{5}$$

Comparing (1) with (5):

$$\begin{cases} -(1+\beta) = -2 \Rightarrow \beta = 1 \\ \alpha - 1 = 0 \Rightarrow \alpha = 1 \end{cases}$$

Choice (2) is the answer.

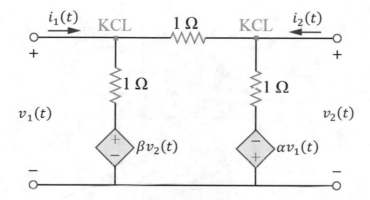

Figure 10.9 The circuit of solution of problem 10.9

10.10. The two networks have been connected in parallel. Therefore:

$$[Y] = [Y_1] + [Y_2] = \begin{bmatrix} 1 & 2 \\ 2 & 3 \end{bmatrix} + \begin{bmatrix} 1 & 2 \\ 2 & 6 \end{bmatrix} = \begin{bmatrix} 2 & 4 \\ 4 & 9 \end{bmatrix}$$

Choice (1) is the answer.

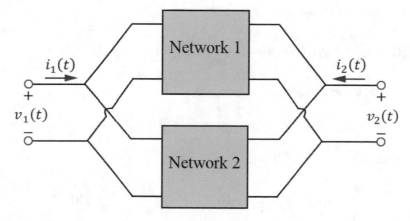

Figure 10.10 The circuit of solution of problem 10.10

10.11. The two networks have been connected in series. Therefore:

$$[Z] = [Z_1] + [Z_2] = \begin{bmatrix} 1 & 2 \\ 2 & 3 \end{bmatrix} + \begin{bmatrix} 5 & 6 \\ 6 & 7 \end{bmatrix} = \begin{bmatrix} 6 & 8 \\ 8 & 10 \end{bmatrix}$$

As we know, the impedance matrix ($[Z]$) of a network is defined as follows:

$$\begin{bmatrix} v_1(t) \\ v_2(t) \end{bmatrix} = \begin{bmatrix} z_{11} & z_{12} \\ z_{21} & z_{22} \end{bmatrix} \begin{bmatrix} i_1(t) \\ i_2(t) \end{bmatrix}$$

Therefore:

$$\begin{bmatrix} v_1(t) \\ v_2(t) \end{bmatrix} = \begin{bmatrix} 6 & 8 \\ 8 & 10 \end{bmatrix} \begin{bmatrix} i_1(t) \\ i_2(t) \end{bmatrix}$$

Choice (4) is the answer.

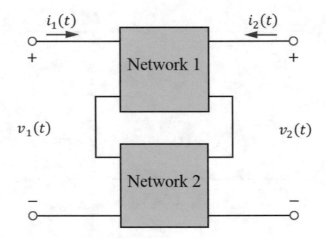

Figure 10.11 The circuit of solution of problem 10.11

10.12. The hybrid matrix ($[H]$) in time domain is as follows:

$$\begin{bmatrix} v_1(t) \\ i_2(t) \end{bmatrix} = \begin{bmatrix} h_{11} & h_{12} \\ h_{21} & h_{22} \end{bmatrix} \begin{bmatrix} i_1(t) \\ v_2(t) \end{bmatrix} \Rightarrow [H] = \begin{bmatrix} h_{11} & h_{12} \\ h_{21} & h_{22} \end{bmatrix} \tag{1}$$

Applying KVL in the indicated loop:

$$-v_1(t) + \beta_2 v_2(t) + v_2(t) - \beta_1 v_1(t) = 0 \Rightarrow (1 + \beta_2)v_2(t) = (1 + \beta_1)v_1(t) = 0 \Rightarrow v_1(t) = \frac{1 + \beta_2}{1 + \beta_1} v_2(t) \tag{2}$$

Applying KCL in the indicated supernode:

$$-i_1(t) + \alpha_1 \widetilde{i_1}(t) + \alpha_2 \widetilde{i_2}(t) - i_2(t) = 0 \Rightarrow i_1(t) + i_2(t) = \alpha_1 \widetilde{i_1}(t) + \alpha_2 \widetilde{i_2}(t) \tag{3}$$

Applying KCL in node A:

$$-i_1(t) + \alpha_1 \widetilde{i_1}(t) + \widetilde{i_1}(t) = 0 \Rightarrow \widetilde{i_1}(t) = \frac{1}{1 + \alpha_1} i_1(t) \tag{4}$$

Applying KCL in node B:

$$-\widetilde{i_2}(t) - \alpha_2 \widetilde{i_2}(t) + i_2(t) = 0 \Rightarrow \widetilde{i_2}(t) = \frac{1}{1 + \alpha_2} i_2(t) \tag{5}$$

Solving (3), (4), and (5):

$$i_1(t) + i_2(t) = \frac{\alpha_1}{1 + \alpha_1} i_1(t) + \frac{\alpha_2}{1 + \alpha_2} i_2(t) \Rightarrow \frac{1}{1 + \alpha_1} i_1(t) + \frac{1}{1 + \alpha_2} i_2(t) = 0 \Rightarrow i_2(t) = -\frac{1 + \alpha_2}{1 + \alpha_1} i_1(t) \tag{6}$$

Solving (1), (2), and (6):

$$\begin{bmatrix} v_1(t) \\ i_2(t) \end{bmatrix} = \begin{bmatrix} 0 & \dfrac{1 + \beta_2}{1 + \beta_1} \\ -\dfrac{1 + \alpha_2}{1 + \alpha_1} & 0 \end{bmatrix} \begin{bmatrix} i_1(t) \\ v_2(t) \end{bmatrix} \Rightarrow [H] = \begin{bmatrix} 0 & \dfrac{1 + \beta_2}{1 + \beta_1} \\ -\dfrac{1 + \alpha_2}{1 + \alpha_1} & 0 \end{bmatrix}$$

Choice (2) is the answer.

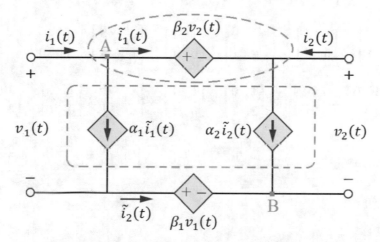

Figure 10.12 The circuit of solution of problem 10.12

10.13. The transmission matrix ($[T]$) of a circuit is defined as follows:

$$\begin{bmatrix} v_1(t) \\ i_1(t) \end{bmatrix} = \begin{bmatrix} t_{11} & t_{12} \\ t_{21} & t_{22} \end{bmatrix} \begin{bmatrix} v_2(t) \\ i_2(t) \end{bmatrix} \Rightarrow [T] = \begin{bmatrix} t_{11} & t_{12} \\ t_{21} & t_{22} \end{bmatrix} \tag{1}$$

The transmission matrix of the circuit shown in Figure 10.13.2 can be determined as follows:

Applying KVL in the loop:

$$-v_1(t) + v_2(t) = 0 \Rightarrow v_1(t) = v_2(t) \tag{2}$$

Applying KVL in the right-side mesh:

$$-v_2(t) + R(i_1(t) - i_2(t)) = 0 \Rightarrow i_1(t) = \frac{1}{R} v_2(t) + i_2(t) \tag{3}$$

Solving (1), (2), and (3):

$$\begin{bmatrix} v_1(t) \\ i_1(t) \end{bmatrix} = \begin{bmatrix} 1 & 0 \\ \dfrac{1}{R} & 1 \end{bmatrix} \begin{bmatrix} v_2(t) \\ i_2(t) \end{bmatrix} \Rightarrow [T_p] = \begin{bmatrix} 1 & 0 \\ \dfrac{1}{R} & 1 \end{bmatrix} \tag{4}$$

The transmission matrix of the circuit shown in Figure 10.13.3 can be determined as follows:

Applying KVL in the loop:

$$-v_1(t) + Ri_2(t) + v_2(t) = 0 \Rightarrow v_1(t) = Ri_2(t) + v_2(t) \tag{5}$$

From the circuit, it is clear that:

$$i_1(t) = i_2(t) \tag{6}$$

Solving (1), (5), and (6):

$$\begin{bmatrix} v_1(t) \\ i_1(t) \end{bmatrix} = \begin{bmatrix} 1 & R \\ 0 & 1 \end{bmatrix} \begin{bmatrix} v_2(t) \\ i_2(t) \end{bmatrix} \Rightarrow [T_s] = \begin{bmatrix} 1 & R \\ 0 & 1 \end{bmatrix} \tag{7}$$

The circuit of Figure 10.13.1 can be assumed like the series connection of seven small circuits. The total transmission matrix of such a circuit can be determined as follows:

$$[T] = [T_{p1}] \times [T_{s1}] \times [T_{p2}] \times [T_{s1}] \times [T_{p2}] \times [T_{s1}] \times [T_{p1}]$$

$$= \begin{bmatrix} 1 & 0 \\ \frac{1}{1} & 1 \end{bmatrix} \times \begin{bmatrix} 1 & 1 \\ 0 & 1 \end{bmatrix} \times \begin{bmatrix} 1 & 0 \\ \frac{1}{0.5} & 1 \end{bmatrix} \times \begin{bmatrix} 1 & 1 \\ 0 & 1 \end{bmatrix} \times \begin{bmatrix} 1 & 0 \\ \frac{1}{0.5} & 1 \end{bmatrix} \times \begin{bmatrix} 1 & 1 \\ 0 & 1 \end{bmatrix} \times \begin{bmatrix} 1 & 0 \\ \frac{1}{1} & 1 \end{bmatrix}$$

$$= \begin{bmatrix} 1 & 0 \\ 1 & 1 \end{bmatrix} \times \begin{bmatrix} 1 & 1 \\ 0 & 1 \end{bmatrix} \times \begin{bmatrix} 1 & 0 \\ 2 & 1 \end{bmatrix} \times \begin{bmatrix} 1 & 1 \\ 0 & 1 \end{bmatrix} \times \begin{bmatrix} 1 & 0 \\ 2 & 1 \end{bmatrix} \times \begin{bmatrix} 1 & 1 \\ 0 & 1 \end{bmatrix} \times \begin{bmatrix} 1 & 0 \\ 1 & 1 \end{bmatrix}$$

$$\Rightarrow [T] = \begin{bmatrix} 26 & 15 \\ 45 & 26 \end{bmatrix}$$

Choice (4) is the answer.

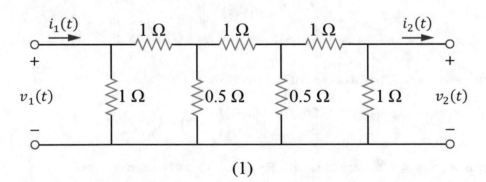

(1)

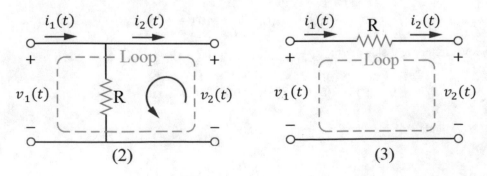

(2)   (3)

Figure 10.13   The circuit of solution of problem 10.13

References

1. Rahmani-Andebili, M. (2020). DC Electrical circuit analysis: Practice problems, methods, and solutions, Springer Nature.
2. Rahmani-Andebili, M. (2020). AC Electrical circuit analysis: Practice problems, methods, and solutions, Springer Nature.

Index